Teaching
Science Skills
at Home

(Without Being a Rocket Scientist)

By Robin J. Schneider

with Karl M. Schneider

Cartoons by Cathy S. Taylor

Library of Congress Control Number: 2005932609

First Printing   August 2005
Second Printing   May 2006
Second Edition   July 2010

Printed in the United States Of America by

Homeschool Zoo

http://www.homeschoolzoo.com

Teaching Science Skills at Home

Table of Contents

Acknowledgements

This book could not have been written without the loving support of my husband, Karl. His many insights and superb editing turned the project from a simple collection of ideas into a real book. In addition to contributing in an intellectual way, he watched the children and ignored my sorry housekeeping skills.

Thanks are also due to my children, who patiently served as guinea pigs for my experiments in various methods of teaching.

The original models for my teaching methods were my wonderful parents. Their innovation in education and patience with their seven children have been remarkable.

Thanks to Cathy Taylor, a most kind and patient illustrator.

Additional thanks to the members of South Bay Free Scholars and the ladies at the Homeschooler's Curriculum Swap, who have asked me so many questions about teaching science over the years that I have been forced to think through and refine my methods. It was this constant barrage of questions which prodded me to write this book.

INTRODUCTION

I must teach my kids science. Must teach science. Must teach science. AAAAAAH!

If you're a homeschool parent, thoughts like these have no doubt run through your head at some point.

Well, why must we teach science? What's the point? "Science" is just a word; to some of us, a big and scary word. That word "science" is responsible for nuclear bombs[1]. We don't want to teach our kids how to blow up a planet (though kids will say that sounds like fun).

So, what goals are we trying to achieve?

GOAL #1: LEARN FACTS

The most obvious goal is to learn facts about the world and to understand how nature operates.

For this reason, many science textbooks focus on rote memorization of science facts. For example, children might memorize the names of planets in the solar system.

There is a problem with the memorization approach; scientific "facts" are subject to change at any time. For example, in the nineteenth century, astronomers viewed the planet Venus through telescopes and—seeing that it was covered in clouds—concluded that it had large amounts of water. The "fact" that water existed on Venus was widely accepted. In the twentieth century, in light of new evidence, scientists concluded that the Venusian clouds contained not water, but deadly sulfuric acid.

Science is a moving vehicle, not a museum showroom frozen in time. The essence of science is participating in moving the vehicle by being an agent of change. Memorization of science facts helps you understand where you starting from. It is the launch pad, not the destination.

Learning facts should not be the only aim of a science curriculum. A good scientist doesn't just know the facts that already have been discovered—he actively seeks to discover new ones.

---

[1] As well as everything from light bulbs to refrigerators to life-saving medicines.

GOAL #2: DO EXPERIMENTS

Here is where the "hands-on" approach to science education comes into play. When children do an experiment, they realize that the rules of the world can be discovered. They realize science facts haven't appeared out of thin air. They actively seek to discover new facts.

Ideally, hands-on science complements (rather than totally replaces) memorization. After all, children can't be expected to discover everything they need to know by themselves; some scientific principles just can't be uncovered using home science equipment. (Anyone want to demonstrate the existence of an atom?)

Many science curricula designed for the home take the hands-on approach. The experiments can be a lot of fun. It's a wonderful feeling when you see your children experience the "light bulb moment" of discovery. Hands-on experiments should be part of every child's science education.

Hands-on curricula have only one real shortcoming, but tragically, it's a big one. They can be so fixated on getting the child to discover that they ignore the third and final aspect of science education: learning Science Skills.

GOAL #3: LEARN SCIENCE SKILLS

Science Skills are far more important to your child than Science Facts. Science Facts can be looked up if they haven't been learned earlier. Science Skills require time to master. It's like the difference between knowing the fact that toothpaste stops cavities versus being good at brushing.

The Science Skills every child needs are:

- Logical Thinking
- Observation
- Classification
- The Scientific Method
- Record Keeping
- Experiment Design

Not many curricula encourage the use of Science Skills, but if you look hard enough, you can find them. What you *won't* find is specific advice on **how to teach** Science Skills.

Suppose you had a child who wanted to become a professional baseball player, but you didn't know anything about the sport. What would you think of a how-to book that explained the rules of the game in great detail but barely mentioned the necessary skills? A child who relied on such a book might become a good umpire, but he would never become proficient at baseball, much less great. Your child doesn't need someone reminding him that he needs to learn how to bat. What he needs is to have his parent give him specific pointers on how to stand, how to hold the bat, how to swing, etc.

Think about science in the same fashion. It is not enough just to learn the rules of the science game. Nor is it enough to be able to talk about the "facts," "players," and "statistics" of the science world. It is up to you, the parent, to teach your children Science **Skills**. You can start as early as birth.

Do you feel overwhelmed yet?

Don't worry, you won't have to swing any bats—or beakers.

I designed this book for homeschooling parents like me. Most activities can be done with no more than two people (you can be the second, if you only have one child). Necessary materials are easily obtained. The concepts are explained in simple terms, so that even if you aren't a rocket scientist, you can teach them to your children. Whether you homeschool your children or teach them science after they return home from a traditional school, you can teach your child science skills.

Question: "My daughter wants to be an artist. She would rather eat broccoli than study science. Does every child need to learn Science Skills?"

The answer is yes.

It's true that not everyone will grow up to be a scientist. However, my own life is a perfect example of why people need to learn science skills.

I was trained as a scientist. However, for many years, I didn't get a chance to use my degree the way most of my college classmates did. Early in life, I decided to become a stay-at-home mom. By the time I graduated from college, I had traded my laboratory apron, scientific calculator, and safety glasses for a kitchen apron, oven timer, and a beautiful baby boy.

At first glance, the professions "scientist" and "stay-at-home mom" might seem to some like oil and water. Did my parents ever wonder if the expense of sending me to college was worth it? I don't know, but I have never regretted the training I received, for it seemed that I ended up using it all the time.

For example, when I was trying to settle on the "perfect" gardening method, I used the steps of the scientific method and the concepts of variables and control groups many, many times.

4

I struggled for a long time with making the perfect homemade whole-grain bread. One day I was trying a new way of doing things and remembered (only after I had set the dough to rise, of course) that I had tried that same "new" idea before. I suddenly realized that my memory couldn't be counted on!

I solved that problem by keeping a lab notebook in the kitchen to keep records of my kitchen experiments. Sometimes my "write-ups" are fancy; sometimes, they're very mundane. One of the briefer entries reads: "Thanksgiving: Used 5 pounds potatoes. Too many leftovers. Next year, try 3 pounds."

These records have been very useful over the years. I'm quite sure I wouldn't have thought of the idea if I hadn't been taught to keep very careful records in my chemistry labs.

Now if I can use my scientific training, even in the humdrum work of a stay-at-home mother, it's likely that anyone can find these skills useful.

Question: "Won't my child pick up these skills elsewhere?"

Not in my experience.

My parents were superb at teaching me three of the Science Skills: observation, classification, and reasoning. That was about as far as their attempts at home science education went. I guess they assumed that I would pick up the other Science Skills in my public school science classes.

They were wrong. I didn't figure this out until I had problems in my first college labs. It was an eye-opening experience for me: something was lacking in traditional science education.

I vowed to keep my children from encountering similar difficulties. As I thought about how I could give them the tools they needed, and as I heard other mothers complain about science curricula, I realized something: I could help children besides my own.

An Outline of the Book

So now you know what you need to do: teach your children science skills. Okay, you say—but how?

That's what the rest of the book is about.

I will start by discussing how to train your child to think logically.

In the next section, I'll get into specifics on how to teach your children basic Science Skills. The skills are explained in the order you should introduce them to your children. Your children should master these before they reach high school.

In part three, I'll discuss topics relating to experimental design. These are more advanced skills and can wait until middle or high school.

Finally, I'll give you a science education plan. The plan is a road map that will show you exactly what you need to do in your home, at each age.

By the time you finish the book, you'll realize something:

You are capable of teaching Science Skills to your children.

PART ONE

Thinking Like a Scientist

CHAPTER ONE
Reasoning and Logic

Why study reasoning and logic? Anyone can make statements about how the world works. However, if these statements are meant to actually portray the world as it is—to promote understanding of how the universe operates—they must be constructed logically.

Teaching logical thinking skills doesn't require a curriculum. You just need to direct your children's development so that they can think for themselves.

At first, this statement seems like an oxymoron. If you help your children to think, how will they ever think for themselves?

The answer is simple. You are not dictating what your children should think; you're merely guiding them along the pathways that lead to correct conclusions.

You can begin teaching your children about sorting while they are toddlers, and progress to patterns when they are preschoolers.

For more advanced skills, wait until age seven or eight. At some point, usually around the ages of seven or eight, children's cognitive development takes a huge leap. They start to become logical creatures. Developmental biologist Jean Piaget called this the beginning of the "Stage of Concrete Operations." Studies of children's brains show a marked change between the ages of six to eight, which seems to corroborate Piaget's beliefs.

As parents, you notice this change in the way your children talk. They start to understand the reasoning between cause and effect. They think of reasons that something might be happening, and there's actually logic behind their conclusions. Of course, they don't yet have the reasoning capability of an adult, so their logic may sometimes be a little faulty. But overall, they begin to have a method to their madness.

Throughout the book, this developmental milestone is referred to as "the age of reason."

SORTING

This skill is the basis not only for patterning, as described below, but also for scientific classification. So long as you keep the categories relatively simple, you can start this skill as early as age two. Most two-year-olds, for example, could sort a pile containing blocks and toy cars into two appropriate piles. A pile containing five different types of toys—blocks, cars, dolls, stuffed animals, and chess pieces, for example—would likely confuse them.

The best way to introduce your children to the concept of sorting is to have them work alongside you. When it's time to clean up a messy room, you can say, "Let's start with all of the books. Here's one book. Can you find more?" Next, "Now let's do the stuffed animals. Here's one stuffed animal. How many more can you find?"

When it's time to put away dishes, have your children sort the silverware into a divided tray: forks in one compartment, spoons in another. (To make things easier and safer for a little one, it's best to remove the knives and put them away on your own.)

When it's time to do laundry, you can sort clothing into shirts, pants, socks, dresses, and so on. Another way to have your child sort the laundry is by owner: Daddy's, Mommy's, older brother's, etc.

Later, your children can learn to sort shapes (triangles, rectangles, squares, circles, etc.) and other, more academic, objects.

PATTERNS

The next foray your children will make into logical thinking skills is learning to recognize patterns. You can begin as early as age three, provided you keep the patterns relatively simple.

10

It's easy to introduce basic pattern concepts using blocks of different colors. For example, you could stack red, blue, red, blue, red, blue, and ask: "What would come next?"

As your children age, you'll want to make the patterns harder and harder to puzzle out.

By the time your children reach the age of reason, you will be able to abandon the building blocks in favor of more abstract items. Finding rhyme schemes in poetry is a fun way to get children to examine things closely in search of a pattern. This is closely linked to the science skill of classification, which will be discussed in more detail later.

Older children who have mastered the ability to detect patterns should be taught to exercise caution. Once the human brain has become adept at finding patterns, it sometimes sees patterns where none in fact exist—the animals in the clouds or the stars, for example, or patterns in the stock market. Professional scientists deploy the power of statistical mathematics to help separate truth from imagination.

 ACTIVITY: Teaching Pattern Recognition

I recommend using blocks when you first start teaching patterns. Any sort of blocks will do—painted wooden blocks, LEGO™ blocks, snap-together Unifix™ cubes, etc. Of course, you don't have to use blocks; anything will work as long as you have a decent quantity. Examples of common household objects you could use include silverware (fork, spoon, fork, spoon…); pasta (macaroni, shell, macaroni, shell…); writing implements (pencil, crayon, pencil, crayon…) and toys (car, doll, car, doll…). I prefer to use blocks because blocks are appealing to children, but not as distracting as some other toys.

First, select a pattern (choose one of the sample patterns on the next page). It's best to start with patterns using just two colors.

Now select one color of block to represent A, one color of block to represent B, and so on. In my examples, I have chosen gray blocks for A and black blocks for B.

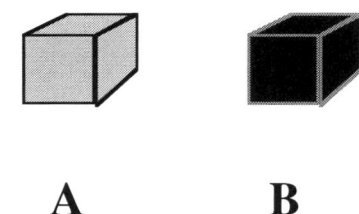

A   **B**

The list entry shows just one "repeat" of the pattern. The younger the child, the more repeats you'll have to show before he spots the pattern. For example, a 3-year-old might need to see "A, B, A, B, A, B, A, B" before selecting A as the next in line. A ten-year-old, on the other hand, would have no trouble with "A, B, A, ___". Decide how many repeats you want of the pattern you want to lay out.

Pattern AB, one repeat

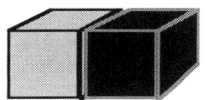

Pattern AB, two repeats

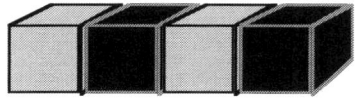

Lay out the pattern on a level surface. For older children, you could simply ask them to complete the pattern. Younger children may need you to read the pattern aloud for them before they are asked to pick what comes next.

🏠 On the Homefront

Most curricula, computer software, and so forth require that your child complete one pattern repeat. That, to me, is a minimum. My own children seem to enjoy repeating patterns until they run out of materials or table space.

Sample two-color patterns
AB
ABB
AABB

Pattern ABB, two repeats

Sample three-color patterns
ABC
ABBC
ABBCC
AABBCC
ABAC

Pattern ABC, two repeats

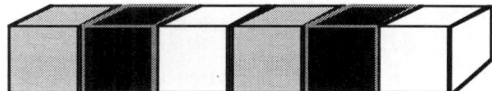

Sample four-color patterns
ABCD
ABBCD
ABBCDD
AABCCD

Pattern ABCD, two repeats

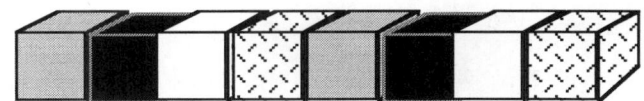

13

RELATIONSHIPS

Patterning skills are related to the skill of recognizing logical relationships. A logical relationship connects two or more objects in some fashion—it is not a random association. Your sister and her boyfriend might have a relationship, but it might not be a logical one!

Actually, a logical relationship can consist of two opposites. "Open window" and "closed window" are diametrically opposed concepts, but inversion is considered a logical relationship. So that sister of yours can—technically speaking—argue that she has a "logical relationship" despite being the complete opposite of her boyfriend.

Those who took the old version of the SAT may remember using this logical skill on those dreaded analogy problems.

Analogies require logical thinking and shouldn't be introduced until your child has reached the age of reason. Even then, it wouldn't be appropriate to introduce an SAT-level analogy to your beginners. They might not comprehend the vocabulary, much less the abstract concepts, in a stumper such as this one:

Articulate: speech :: coordinated : _____.

High school students should be able to figure out that the missing word in this analogy is "movement." Someone who is articulate is good with speech, just as someone who is coordinated is good with movement.

Instead of using abstract concepts with your beginners, try analogies which use concrete objects. Here's a simple analogy that you can use with a beginner:

Fingers are to hand as toes are to _____.

When you start doing problems of this type with a child, begin with oral practice. The practice sentence should be read like this: "Fingers are to hand as toes are to WHAT?"[2]

To an adult, it's obvious that the answer is "foot." But what if your child can't figure this out?

In this example, you could ask: "How are fingers related to hands?" Any number of replies are possible, but the gist of the answer should be that fingers are body parts which are attached to a hand.

---

[2] The sample SAT analogy given above is read in the same fashion. The colons are shorthand for "is to", while the double colon represents "as".

Now that you've established the first relationship, you can work on finding the missing part of the second relationship. "Fingers are body parts which are attached to a hand. Toes are body parts which are attached to what?"

Once your child figures out the answer, you can reinforce the concepts by repeating the relationship: "That's right. Fingers are attached to hands in the same way that toes are attached to feet."

CAUSE AND EFFECT

Getting your child to learn about cause and effect is a long process which is sure to try the patience of any parent.

Part of the problem is that very young children don't have the reasoning capacity to make proper connections between events. Sometimes they will see two events and not realize that one causes the other; at other times, they'll see two events and think they're related when, in fact, they're not.

And yet, despite their lack of cognitive development in this area, young children naturally ask "Why?" In fact, at some ages, it might be more accurate to say that they ask, "Why? Why? Why? Why? Why?" The key is to take advantage of this natural curiosity and answer each and every "why."

On the Homefront

Many children delight in burying you with questions. I remember one of my children approaching me with a long string of "why" questions—"Why this?" "Why that?" "Why are you so sure?" When he sensed that I was near the end of my rope, he concluded with a parting shot: "And why do I always ask why?"

Do your best to answer each query with a reasonable answer. Your consistent answers encourage their curiosity; your explanations help them to develop their reasoning skills.

When your children are preschoolers, give them a simple explanation every time they ask. You need to walk the fine line between not giving them enough information and overwhelming them. That may sound difficult, but it becomes easier with a little practice.

15

As they approach school age, you can start to expand this somewhat. You can explain why things are that way, and then tell how we can learn that.

Sometimes your answer will spawn another question for you to tackle; sometimes, a whole chain of questions will result! This can be aggravating to a busy mother, but do try to be as patient as possible in answering questions. This is one of the best things you can do to develop a child's intelligence.

At times, you will not know the answer. Show your children where you can look to find it. Hopefully you have some sort of reference around your house that you can turn to. If not, check your local library or the Internet for answers.

🏠 On the Homefront

Thanks to a typo, a friend of mine, along with her children, once stumbled into an undesirable corner of the Internet. Like her, my home internet connection is unfiltered, and my children haven't mastered spelling. I now type in the search myself, and then call my children over to view the results.

By the time your children reach the age of reason, they are ready to have the tables turned on them: you start answering their questions with questions! You aren't doing this to be deliberately annoying; you pick your questions specifically so that the answers give your children clues. You will be using the ancient art of teaching through questions—often referred to as "The Socratic Method"—which will be explained in more detail in the next chapter.

IN SUMMARY

Give your younger children a gentle introduction to the art of reasoning. Teach them to recognize patterns and relationships. Embrace their enthusiasm and find the energy to answer their questions, to feed their hunger for knowledge. By the time they reach the age of reason, they should be ready for learning with the Socratic method.

CHAPTER TWO
Teaching with the Socratic Method

"Mommy, can I have more ice cream? Where does ice cream come from?"

As a parent, you're accustomed to hearing all kinds of questions from your children; generally, you respond immediately.

While conversations that follow this pattern can be enlightening, they aren't necessarily the best way to teach your children. If you want your children to really understand the whys and hows of life, use the Socratic method in your conversations.

 On the Homefront

What does a Socratic Method conversation sound like in real life? Here's a conversation I had with one of my boys, then age eight, when we were out for a walk. This gives you an idea of what the Socratic Method is about.

> Son: "Mom, why are there those little holes in that tree?"
> Mom: "Do you have any guesses about what made them?"
> Son: "Well, at first I thought it was a woodpecker, but those are way too small for a woodpecker-house."
> Mom: "A woodpecker is a good guess. Suppose it *was* a woodpecker that made those holes. Why else would he peck at a tree? Can you think of a reason besides making a house?"
> Son: "Maybe looking for something?"
> Mom: "Something? Like what?"
> Son: "Like maybe food?"
> Mom: "Well you're right. It was a woodpecker looking for food. Do you know what sort of food he was looking for in the tree?"
> Son: "Bugs, I think."
> Mom: "Right!"

My natural conversational style tends towards constant lecturing. Although it was difficult for me to transition from lengthy discourses to the more interactive conversation of Socratic dialogue shown above, I've found it worthwhile to exercise their brains.

The Socratic method is named after one of the greatest philosophers of all time (Socrates). If you think yourself incapable of following in his storied footsteps, take heart. The Socratic Method is not rocket science – after all, it was invented in the fifth century B.C.

The underlying principle of the Socratic Method is that you don't give a straight answer to a question. Instead, you ask questions which help your children to figure out the answer themselves.

Of course, when it comes to some questions—like "Mommy, can I have more ice cream?"—maybe the Socratic Method isn't a parent's best choice. But when it comes to "where does ice cream come from?" and many other questions, the Socratic Method is the way to go.

FOUR STEPS TO SOCRATES

Now that you've seen an example of the Socratic method in action, let's try to extract what it is that you are doing or supposed to be doing. The following four steps are one way to achieve Socratic dialogue.

(1) Ask questions about things your children can directly observe.

Example:
Suppose your child asks you, "What color will the baby's eyes be after it's born? Do you think they'll be blue?" Obviously, there's no way to observe baby's eyes when it's still *in utero*, but there is something you can observe right now. Ask:

- "What color are Mommy's eyes?"
- "What color are Daddy's eyes?"
- "How about your eyes?"
- "What about your siblings—what colors are their eyes?"

We will return to this example in the next step.

Comments:
It's ideal to start with direct observations because they are indisputable. If I think it is sunny today but you think it's cloudy, you and I can go outside together to find out the truth. If your children complain that their toast looks like charcoal, it is possible to verify that fact.

Of course, direct observations are not always possible. If your children have a question about the workings of the liver, how could you possibly observe a live liver? And if they ask you about black holes—well, forget it. So don't worry if you sometimes skip this first step.

(2) Ask questions about facts your children are already familiar with.

Example:
Let's return to the question about the baby's eye color. Your child has observed the eye colors of all family members, but still doesn't have enough information to come to a conclusion. In order to understand the situation properly, he needs to be reminded of basic principles of genetics. Here are some questions you could ask:

- "What decides how a baby will look?" (Many children won't know the answer to this right away; you may have to treat it as a rhetorical question and help them deduce the answer with the questions that follow.)
- "Do you remember where babies come from?"[3]
- "What fraction of the information about eye color does a baby get from its father? What fraction of the information about eye color does a baby get from its mother?"
- "Do you remember what we call the pieces of information that a baby gets from its parents?"
- "If a baby gets one gene for blue eyes and one gene for brown eyes, what color will its eyes be?"

Comments:
In steps one and two, you are building a foundation of facts, which you will use to reach your final answer. The purpose of this phase of questioning is to introduce facts which are *not* directly observable.

If you've elicited enough direct observations from your children in step one, skip this step. Direct observations are best. Why? Quite frankly, some "facts" your children know are less than factual. Children tend to latch on to anything they hear. For years my little sister thought that chocolate milk came from dirty cows!

(3) Ask questions that help children draw conclusions from the facts and observations you've just reviewed.

Example:
Now that your children have been reminded of basic genetics, you need to get them to draw conclusions. Here are some questions you might ask:

---

[3] "Where do babies come from?" is a question that strikes fear into the heart of most parents. The explanation I gave my children when they were young was, "The daddy gives a piece and the mommy gives a piece, and when the two pieces join they can make a baby." This tidbit of information was enough to satisfy my children until age nine or ten, yet was useful enough that they could understand elementary genetics.

- "What eye color genes do you think Mom has to give to the baby?"
- "What eye color genes do you think Dad has to give to the baby?"
- "What is the likelihood that both Mom and Dad will both give a blue eyed-gene to the baby?"

<u>Comment:</u>
In the first two steps, your children gathered all the necessary data for getting to the final answer. Now, they need to take that data, process it, and figure out how it relates to the situation at hand.

Keep asking them to draw conclusions until they are just a hair's breadth away from the final answer.

(4) Repeat the original question.

<u>Example:</u>
- "So, do *you* think that the baby will have blue eyes?"

<u>Comment:</u>
If your children can't answer the original question at this point, don't panic. It just means that you need to go back to step #3 for a while. Then let them take another stab at it.

Remember, the solution is obvious to you. You know where all these questions are leading; they don't. Your questions may have stopped far short of the mark. When you were thinking through the logic, you may have taken a huge leap forward. Since you already know the answer, this jump seems trivial to you, but it is a huge hindrance to them.

Usually, given enough encouragement and hints, children of this age *can* puzzle out the answer. With enough practice, they will learn to think through problems in a logical and orderly fashion.

For more examples of using the Socratic Method, please see Appendix I.

COMMON QUESTIONS

Do I have to use the Socratic Method every time my child asks a question?

When your child asks whether or not he can go to a friend's house to play, you will just say "yes" or "no." If the question is a factual question, however, it is best to start with the Socratic Method.

While Socratic dialogue may be unfamiliar and difficult at first, it is the best tool available for teaching logical thinking. It also turns many everyday moments into learning opportunities. You'll also find that the Socratic method gives you lots of bonding time with your children.

The ebb and flow of natural conversation doesn't always adhere to the four-step plan. Isn't it too rigid?

It is hard to give an exact formula for using the Socratic Method. The questions you ask your children will depend on the original question that you were asked. I outlined four basic steps to get you going, but keep in mind that you need to be flexible. You may end up not following this process exactly.

I'm still not sure what questions I should ask in the first three steps. Can you give some more examples?

In Step 1, you are trying to encourage direct observations, so ask questions related to the senses (sight, smell, touch, hearing, taste). Rather than making generic inquiries ("What do you see?") it's better to be specific. For example, the following questions all require using the sense of sight to find the answer:

- "What is that bird doing?"
- "Which plant is taller?"
- "What shapes are the rocks at the bottom of the stream?"
- "Are all the apples the same color?"

Before you ask questions in Step 2, take a moment to think of the facts your children need to get to the final answer, and ask questions to get your children to recall those facts. For example, if answering the question requires an understanding of the digestive tract, you might ask:

- "Where does food go when you swallow it?"
- "Where does it go after the esophagus?"
- "What happens to the food in the stomach?"

In Step 3, you need to ask questions that force your children to draw conclusions about the facts they've gathered in the first two steps. For example:

- "This phenomenon occurs during winter, but not during summer. What does this tell us about what makes it happen?"
- "This is true in one situation. Do you think it would be true in all situations?"
- "We know what happens if X occurs. We know what happens when Y occurs. What do you think would happen if they happened at the same time?"

What do I do if my child loses patience?

Pay careful attention to your child as you ask questions. If he seems to be losing patience, summarize what he has figured out so far and then explain the rest of the conclusions. There is no sense in making a child frustrated.

As you practice Socratic dialogue, your child's tolerance for figuring out answers will grow. It may be a year or more before your child can last through all four phases of questioning. Be patient! Your child's reasoning skills will mature with time.

I've gone through all four steps. My kids still don't get it. Am I a failure?

No, of course not! If you're new to this method of teaching, your first reaction is that you are doing it wrong. And while you may need to gain experience, don't put all the blame on yourself. It's not uncommon for children to have trouble with Socratic dialogue when they're first introduced to it. For example, I like to think that I'm pretty good at teaching children this way, yet every time I try it on a new tutor client or student, I have to back off and "take it easy."

Even when your children become familiar with Socratic teaching, there will be times when they get stumped, times when they just don't see the things you want them to observe. If different methods of prompting don't help them notice appropriate facts, point out the relevant observations yourself. Likewise, if they don't remember (or haven't yet learned) the facts they need to draw appropriate conclusions, explain what they need to know before asking them what those facts might imply.

Even if you make all the observations for them, relate all the relevant facts, and draw the appropriate conclusions, you are still not a failure. You will have imparted far more information to your children than you would have with a simple answer. Furthermore, you will have shown them logical thought in action.

Over time, your children will become accustomed to the notion of Socratic dialogue, and soon they will be eager and willing participants in your conversation.

In Summary

The Socratic method teaches children to examine the evidence available and draw logical conclusions. Once they understand the process, they will know the correct procedure for deducing answers, and may only need minimal prompting to solve mysteries by themselves. Socratic dialogue also prepares children for critical thinking, an advanced skill that they will learn during their high school and college years.

PART TWO

Basic Science Skills

Chapter Three
Observation: The Key to Understanding the World

Observation is one of the most fundamental skills that a budding scientist can learn. It is the starting point for all scientific inquiry. Unless you are a careful observer, you will never ask, "Why are things the way they are?"—you won't really notice the way they are in the first place!

Luckily, this very important skill is one you can encourage and develop in your children very early, as early as the toddler years.

Introductory Observation

Begin by teaching your children to observe *objects*. Generally speaking, this is best done through nature study. In the early stages, you won't be learning to classify plants or anything like that; you'll just be looking very closely at what you see. Nor will you need to take any special books or equipment with you (unless you want to bring a magnifying glass); just go on walks and talk about what you find.

On the Homefront

Nature walks have been an integral part of our family's science study for years. Sometimes I take my boys out to an open space preserve or state park for these walks. Often, we just walk around the block and look at trees, flowers, or other plants that are found in our neighborhood. Once we looked only at weeds!

If possible, grow a garden as well and have your children help. If you live in an apartment, this could be a container garden; it doesn't matter so much where you grow things as long as you have growing, changing plants available for your children's observation.

From the point of view of education, growing edible plants is preferable to flowers or other ornamental plants; the act of harvesting forces your children to get out and look at the plants on a regular basis. Most of us "city slickers" really don't have much experience with where our food comes from, so planting a garden is doubly educational.

It's amazing to put a little seed in the ground and then, later, see the little seedling push up. It's fun to get out and observe the little flowers, then come back later and see the pea (or tomato or whatever) forming where the flower was. It's neat to leave the broccoli on the plant just a little too long and watch the buds swell and then burst into bloom.

OBSERVATION WITH YOUNG CHILDREN

When your children are not yet verbal, you get to do the observation for them—out loud, of course. A toddler has a short attention span, so you might want to start by limiting your observations at that stage to one or two per item.

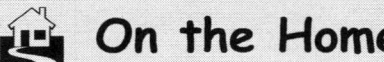

On the Homefront

Teaching a baby or toddler to observe requires a great deal of concerted effort. Following in my mother's footsteps, I always tried to open the eyes of my very young children to new ways of looking at the world.

For example, I would tell a one-year-old:

"Look at this big tree—see how tall it is?" As I said this, I would point upward in an effort to get him to take in the height of the tree.

"Feel the bark. It is very rough, isn't it?" My boys especially enjoyed touching different textures at a very early age. Observation may be done with the hands as well as the eyes.

Many people, at this age, limit their observations to obvious details like color, size, and shape. Try expanding your children's horizons by exposing them to other categories of description. See the section on description categories, Appendix I, for ideas.

When you're talking to a young child like this, you'll want to be at his level. Carry him around while you're talking, or squat or kneel next to him when he's standing by himself.

By age three or so, you can involve children in the observation process by asking simple questions. A simple question requires only one or two words for an answer.

(It's quite possible that some children will be speaking in sentences at this age, but they probably aren't ready to answer essay questions yet.) It's all right to ask more than one question about a single object as long as you remember that children don't have a terribly long attention span at this age.

"This is a very pretty flower. What colors do you see? These pink parts on the outside are the petals. Shall we count the petals together?"

If your children enjoy the counting process, you can continue along the same lines. "This flower has five petals. Here's another flower that looks just the same. Shall we count the petals on this one too?" You might even add, "I wonder if all of these pink flowers have five petals?"

OBSERVATION IN THE ELEMENTARY GRADES

By the time children reach kindergarten, they are ready to give their observations in the form of sentences. Most children are natural collectors. My boys seem to gravitate towards rocks, though they've gone through stages of bringing home leaves, flowers, and even sticks.

The price to be paid for bringing one of these "treasures" home is to have to describe it to a parent. Ideally, you copy these dictations into a nature journal.

This description was written by a child in early elementary grades. Other than being prompted to observe texture, he came up with the description on his own.

This leaf has one big line going through the center and many lines going diagonally. It is oval and green. It is a little bigger than my hand. It smells weird. It feels soft.

For the first few weeks or months, let your children describe the object in any way they want. As they becomes comfortable with this format, you can start stretching their boundaries a little. Always let them describe as much as possible before you start prompting. However, if it's clear that they are finished, ask them a question about a description category they have missed.

Say, for example, that a child is describing a rock. The child tells you that it's very smooth (texture), oval (shape), and gray with black speckles (color). You might try asking, "How would you describe the size?" or, "How heavy do you think this is?"

Objects Children Enjoy Observing

Rocks. These are listed first because they're easiest for a novice to describe. Living things are much more complex in appearance.

Seashells. Those who can make it to the beach will probably find these easier to describe than rocks, making them ideal for early observations.

Flowers. Children are usually interested in the flower itself. Older children should be taught to observe the stem and leaves as well.

Trees. Children focus primarily on leaves and occasionally observe cones or seed pods as well. Make sure they also look at the bark and branching patterns.

Birds. Wildlife that you can watch from your window! Put out a bird feeder in an area that's easily visible from your school area (or wherever you spend most of your time during the day). Even in a fairly urban area, you should be able to see two or three types of birds each day. In the suburbs, you may see half-a-dozen or more species on a regular basis. The actual number will depend greatly on what sort of landscaping you have around your house.

Bugs. This is a love-it-or-hate-it category. If you have children who like bugs, buy some bug boxes and practice catch-and-release; for mom's sanity, bugs should always be sent back into the wild!

At some point your children will be writing fluently enough that they can write these descriptions themselves. When you make the switch to child-written descriptions, you will need a notebook that can serve as a nature journal.

Those who can draw may enjoy doing so. However, don't let the thrill of pictures eclipse the importance of writing. Adults communicate important information primarily through the written word, so it's vital that your children learn to do so as well.

Once your children are comfortable with written descriptions, it's time to move on. You can, of course, continue the exercises that follow with dictated descriptions, but it may be easier to wait until they can write decently on their own before continuing.

METHODICAL OBSERVATION

The next phase is to introduce your children to the physical description categories described on pages 128-129. Children are rarely methodical enough to do this before age 7 or 8, and some may have to wait until they're even older for this step.

Before you introduce the concept, you will need to have prepared a list of categories printed neatly and in large type. A photocopy of one of the lists in Appendix V will do, or you can make your own. You may want to laminate this or put it in a sheet protector, so you can mark on it with a wipe-off pen or pencil. Keep this list hidden for now.

Once again, you have your children describe an object to the best of his ability (either dictating to you—in which case you <u>must</u> write down exactly what they say—or writing by themselves). When you're done, take out the list you've prepared and show it to them. If they aren't reading well on their own yet, you can read the words aloud to them, explaining what they mean, if necessary.

Return to the dictated description and read the first characteristic. Ask your children which category they think this belongs in. (If they are pre-readers, read through the categories again, telling them to stop you when they hear the right answer.) Make a little checkmark next to the item on the list so you know that it's been covered. When you've gone through the whole description in this manner, look at the list of categories again and ask which have been missed. Of course, if you're using an object picked up outside, make it clear that it might not be a good idea to taste the item. Maybe they would like to describe something next time where tasting is okay (a piece of fruit, perhaps, or maybe just their lunch).

Do this exercise with his descriptions over the course of several weeks. Then you are ready to take the next big leap: see if they can, on their own, describe an object using all of the categories. Don't prompt them, but do allow them to use the list. Once they say they are done with their description, go through the list category by category. Which descriptive sentence or phrase corresponds with the category of texture? Which matches the category of color? Go on in this fashion until you've covered the whole list.

Once your children have become accustomed to describing objects using all of the description categories, let them return to describing things without the help of the list as a crutch. Hopefully, by this point, they have expanded their observational abilities to the point that their descriptions will be precise and more or less complete.

If your child is having problems, be sure to try the Description Game at the end of this chapter. It's fun!

🏠 On the Homefront

Some might ask, "Must my children use every single one of these categories every single time that they describe something?" Of course not! The problem is that most children tend to have one or two favorite categories that they'll turn to time and time again. They'll ignore the others if they're not trained to look for them.

For example, I have one son who favors quantitative measurements. Early one spring, I sent him out to make a nature journal entry describing our cherry tree. He measured the height of the tree, the diameter of the trunk, and the length of every single branch. He thought he was being thorough. However, it wasn't until I prodded him to describe what was on the branches that he realized that there were buds there!

His brother, describing the same tree on the same day, noticed primarily the colors on the tree. I hadn't noticed before that the bark on the tree changes color as it ages. My son didn't realize the reason for what he was observing, but he noted carefully that the bark further down the tree was grayer than the bark near the top. He didn't write much about size—he added one sentence after he spied his brother measuring—and he completely ignored the texture of the bark.

32

OBSERVING CHANGE

When your children get better at describing objects, they'll be ready for the next level—describing change. Find an object that changes over a period of time (see box on page 34). Observe it at regular intervals, describing it in writing each time. Be sure to date each entry in your nature journal. The speed at which something changes is very important, and the date serves as an important clue.

If you're observing something which is one of many, mark the object you're observing. For example, to distinguish one flower from others on the bush, loosely tie a piece of yarn around the stem.

Slow-Speed Observations

Here are some objects which change relatively slowly.

Observe weekly:
- Deciduous tree in fall (or spring).
- The leftover food in the refrigerator that no one wants to eat.

Observe once or twice a week:
- A seed growing into a plant (you can plant this outdoors or indoors).
- Young animals: kittens, puppies, chicks, rabbits, tadpoles, etc.

Observe daily:
- A rose changing from bud to bloom and beyond (watch until a few days after the petals fall off so your children can observe rose hip formation).
- A vegetable forming (begin at flower stage and continuing until it is harvest ready). Summer squash such as zucchini are the quickest to grow, but any above-ground vegetable can be observed on a daily basis.
- Sprouting seeds. Place a teaspoon or two of alfalfa, mung bean, or wheat seeds in a jar. Cover the opening with cheesecloth, mesh, or other material which will allow air and moisture to pass through. Each day, add enough water to the jar to cover all seeds, swish around, then drain completely. Place jar in a dark area. Observe every day for a week or so. Alfalfa and wheat sprouts should be left in the sunlight for the last day or two to "green up". Sprouts can be added to your salad or sandwich after observations are complete.

Once your children have learned to describe these slow changes over time, they'll be ready to describe processes that happen more quickly. A good warm-up lesson is to watch a pot of water on the stove as it goes from cold to boiling. This seems like a slow process; describing it as it happens helps to pass the time. Contrary to the old maxim, a watched pot will boil!

The next step is to do some simple science experiments and have your children describe what happens. If you need ideas for what to do, try some of the books suggested in Appendix IV.

ACTIVITY: Description Game

If you have a child who seems to have trouble describing something—either orally or in writing—you may want to play the following "description game."

Number of Players: 2
Minimum Age: 8-10 years old
Materials Needed: drawing paper, lined paper (if you're using this as a
                          written, rather than oral, activity), and pencils

Instructions for developing written skills:

First, each person uses drawing paper to draw a picture. Make sure your partner doesn't see it!

What should you draw? Drawing an object is best for those playing the game for the first time; see if the object can be described without using its name. Avoid choosing an object that is visible from where you are sitting, or your partner will be able to draw it easily, even without a good description! Drawing faces, landscapes or other complex pictures makes the game more challenging.

The next step is to write a description of your picture on a sheet of lined paper. Describe every last detail.

When both players have done this, they pass only the written description to their partner. The partner now tries to recreate the picture based on the written description. When finished, compare your partner's version of your picture to the original. What is different? Could you have added words or statements to make your picture more accurate?

Instructions for developing oral descriptive skills:

Start as you do for the written game. Each person draws a picture, which he keeps hidden from his partner. Now take turns: one person describes his picture orally, while the other tries to draw the picture based on the description he hears.

Compare the pictures and discuss as you do for the written version of the game.

🏠 On the Homefront

The first time I played this game with my oldest son, he drew a picture of a battle scene. He wrote, "Something red is lying on its side with an arrow sticking out of it." In his picture, the "something red" was a red dragon defeated in battle. Since I had only the words "something red" to go by, though—and since I wanted to make a point—I drew it as an apple! He was taken by surprise, but he understood exactly what I was trying to say about the vagueness of his description.

CHAPTER FOUR
Classification: Putting Everything In Its Place

Classification helps us make sense of our observations. If you can place something into a group of like objects, you will know a lot more about it.

For example, you see a bird that you've never seen before. You don't know what it's called. However, you could safely assume that, like other birds, it lays eggs, is warm-blooded, and has a beak, wings, and hollow bones.

People often make the mistake of thinking that classification skills are used only in biology for the taxonomy of living things. Actually, there are examples in all branches of science. In earth science, we often classify rocks, minerals, landforms, and clouds. In chemistry, we classify chemicals by their functional groups, properties, and/or position on the periodic table. In physics, we classify sub-atomic particles, simple machines, and so on.

As mentioned in chapter one, very young children can learn a lot about sorting—the prerequisite for scientific classification—by working alongside their parents. When you pick up toys together, you can sort all the LEGO blocks in one bin, all the wooden railway pieces in another bin, and so on. When it's time to put away dishes, little ones enjoy putting plates in one area, bowls in another, and cups in a third.

BEGINNING FORMAL CLASSIFICATION

Children should be ready to begin learning about scientific classification about age seven or eight. See below for activities on classifying vegetables, cars, and airplanes.

Why should you pick vegetables, cars, and airplanes, rather than some other object, such as dogs? What makes the ideal object?

First, the perfect item is something easily found. Children, especially younger children, don't want to move long distances to find things, nor do they want to wait around for long periods of time. Bird watching can be problematic because of these issues.

Noah encounters armadillos for the first time

Second, the optimal subject of study is something ordinary (like cars) that the children have seen before. The goal here isn't to introduce the children to something new and exotic—the goal is to practice classification. Using novel objects will distract children from the point of the lesson.

Third, the ideal object allows you ten or twenty different kinds to choose from. Goldfish are ordinary enough, yet they don't work well because it's hard to find more than a handful of different types in any single location. Even the pet store doesn't carry all that many. There are dogs in every neighborhood in America, yet unless you can visit a kennel, you aren't going to find enough variety for the purposes of classification.

Finally, remember that the ideal object is something the parent has personal enthusiasm for. Your excitement will be infectious, and what might have been a boring lesson will become an adventure to your children.

When you're teaching about classification, is it okay to use photographs of objects?

For the first few lessons, use objects children can actually handle, such as vegetables. Younger children find it more difficult to classify photographs of objects, since the items in the pictures can't be touched, smelled, or rotated.

🏠 On the Homefront

My father was deeply involved in my own early experiences with classification. He was a big fan of airplanes, and as a result he'd often take us down to one of the local airports to go plane watching.

It wasn't until I was older that I realized how much my classification skills had been shaped by those outings. Learning that each type of airplane had certain distinguishing characteristics that set it apart from other types; learning to look for the small details that separated one class of airplane from another, very similar class; and learning that, sometimes, two airplanes can be almost exactly the same despite external coloring were all lessons that were easily transferable to natural objects such as birds or wildflowers.

Even now that I am older and understand the educational value of what I am doing, I still enjoy watching planes take off and land at the airport. It brings back memories of good times with my Dad.

ACTIVITY: Vegetable Classification

A good first foray into classification is to classify vegetables. These have the advantage of being easy to manipulate and rearrange. Unless you keep a large variety of vegetables in your refrigerator, take a "field trip" to a supermarket produce department. You can just look at the displays, but if you buy the vegetables, you can bring them home, cut them open, group and regroup them on the kitchen table, and even taste them.

Set a medley of vegetables on your kitchen tables in random order. The first thing you need to ask your children is, "What are some ways we could sort these vegetables into groups?" Examples would be sorting by size, color, shape, taste, or by what part of the plant they come from. See how many your children can come up with on their own before giving them hints.

The second thing to ask your children is, "What is the *best* way to group these vegetables?" The goal is to have as many like characteristics as possible in your groups. Your grocery store already sorts fresh produce to a limited extent; you don't find onions sitting in between the peaches and nectarines, for example. But you are trying to do a more scientific classification.

To compare classification methods, pick examples from the harvest of vegetables before you. For example, for classification method "size," you could ask, "What does a three-inch-diameter white potato have more in common with: a three-inch-diameter tomato or a five-inch-diameter white potato?"

The answer is the five-inch-diameter potato. Even though the three-inch-diameter potato and the tomato are in the same "size group," they just don't have as much in common as two potatoes do. Size is a good way to classify clothing, but it's a lousy way to classify vegetables.

For classification method "color," try asking, "What does a white potato have more in common with: a white onion or a yellow potato? What does the answer tell you about the importance of color?"

For the classification method "part of plant," you could ask, "What does a red beet have more in common with: a red tomato or a orange carrot? Which is more important as a classification method, color or which part of the plant the vegetable comes from?"

Continue in this fashion until you have examined all the characteristics your children came up with at the beginning of the activity.

ACTIVITY: Car Classification

If you live in an urban area, practice classification using cars. A walk down your street (or through the parking lot of your apartment complex) will yield a good selection of automobiles. If you live in a rural area, you may have to make a "field trip" into town to find a parking lot. The grocery store works, and lets a busy mother kill two birds with one stone.

There are many ways to classify cars. How many can your children come up with? As with vegetables, color is an obvious attribute. Another obvious characteristic is the manufacturer (Ford, Honda, etc.). After that the differences get more and more subtle. See how many your children can come up with on their own before giving them hints.

Of course, the questions you ask depend on the types of cars you see, but they might include:

"Which have more in common: a blue Ford Taurus and a blue Toyota Camry, or a blue Ford Taurus and a white Ford Taurus?"

Given the answer to this question, which do you think is a better classification method– color or model?

"Which set has more in common: a Ford Taurus (sedan) and a Ford Windstar (minivan)? Or a Ford Windstar (minivan) and a Honda Odyssey (minivan)?"

Given the answer to this question, which do you think is a better classification method– brand name or vehicle class?

ACTIVITY: Airplane Classification

If you have a child who's intrigued by airplanes—or just traveling in general—this is a great way to learn about classification. If you don't know the different types of airplanes yourself, pick up a guidebook at the library or search online for aviation photo galleries. Teach your children different engine configurations and tail types, and soon they will be picking out the most common airliners or military jets in your region. Boys in particular might have an interest in learning about military jets.

Sadly, since September 11, 2001, it may no longer be possible to go to the airport just for the joy of watching airplanes. However, if you find a good spot outside the airport perimeter, you can still see the airplanes fairly soon after takeoff (or right before landing).

CLASSIFICATION IN NATURE

As your children get older, you will want to introduce the idea of classification into your nature study. Get a field guide or two and try to classify the things that you see. At first, focus on only one type of object per nature walk. You might learn about wildflowers during early spring, trees a month or two later, and birds a couple of months after that. Try not to overload children with too much information at once.

 On the Homefront

I chose plants to be the subject of my first classification efforts during nature walks because they wouldn't run away from my children (though, given the destructive tendencies of my boys, I sometimes think they would have run away if they could). Birds were fascinating but very frustrating to my younger children. They often flew away before we could look them up in a guide, and we certainly couldn't approach them for a closer look.

Zoo animals are trapped; there is no way for them to escape. Nevertheless, I don't recommend a zoo trip for teaching classification. My experience is that the animals are too far apart—you can't have them stand next to each other for comparison purposes.

Likewise, rocks and minerals aren't going to run away from you. However, it's very difficult to find more than half a dozen different sorts of minerals on a short walk. These are best studied at home with a collection—either self-collected or purchased from a science supply company.

As you classify objects during nature study, remember that the goal is *not* to memorize the names of the plants or other objects—though, admittedly, memorization will occur with repeated exposure to the same objects. The goal is to understand *why* this plant is classified the way it is.

For example, why are the Coast Live Oak and the California Black Oak grouped together in the same genus? What characteristics do they share?

If they share so many common characteristics, why aren't they considered the same species? What differences are there?

If you're doing this out in the wild and can find a specimen of each type of tree, you can compare leaves, acorns, bark, and so on. Doing an in-depth study of two types of trees like this will ingrain their names in your child's memory far more than a simple, "Oh yeah, this one's a California Black Oak."

For those confined indoors during the wintertime, a good project might be classification of animals. Gather a bunch of library books with information on the taxonomy of various species. Write the names of the animals on slips of paper (or, better yet, find pictures in magazines or on the Internet). Then, ask your children to sort them into groups of related animals. When the sorting is complete, you can ask why they sorted the animals as they did.

For younger children, try a general collection of animals which includes reptiles, fish, amphibians, mammals, birds, and so forth. The differences between these classes of animals are large enough that younger children should be able to handle the sorting.

Older children might be challenged by a collection like this: ferret, lion, cow, fox, tiger, buffalo, skunk, wolf, wolverine. All of these animals are from the same class (mammals). Children will have to pay attention to detailed characteristics to classify them properly[4].

Sometimes, at the end, when you and your child go over the classification scheme she's come up with, you'll notice that her idea of grouping is different from that accepted by the scientific community. Why did she group animals the way she did? Why do scientists group the same animals the way they do?

---

[4] Taxonomists group them this way: ferret, skunk, and wolverine (weasel family); lion and tiger (cat family), cow and buffalo; fox and wolf (dog family).

IN SUMMARY

Far more than a tool for taxonomists, classification is one of the major science skills and is applicable to many areas of science. The best news is that it's easy to teach classification, even to younger children. There are an abundance of suitable everyday items, from vegetables to airplanes, and it's a natural fit to nature study.

Classification can be useful when forming hypotheses—for if you know how one thing behaves, it's easier to guess how a similar thing might act. Forming hypotheses is an essential part of the Scientific Method, which we'll discuss in the next chapter.

CHAPTER FIVE

The Scientific Method Explained

"The Scientific Method" is a term that you will hear a lot in science classes. That lovely word "the" at the beginning of the phrase makes it sound like there is one and only one scientific method, like an official law whose text has been certified and approved.

That's not quite the case; you will likely encounter different versions as you go through life. Fortunately, you will find that most of the steps are the same, with only a few minor variations. For example, some people combine two steps into one step for the sake of simplicity.

What matters is that the scientific method provides a methodical way to discover why certain things happen. Not only is the scientific method objective, reproducible, and logical, it is the only method that other scientists will accept.

Here are the steps:

First, **state the question or problem**. Before you go on your voyage of discovery, you need to decide what you're trying to discover!

Second, **do research**. Use reference materials to help you find clues that will help you understand your experiment.

Third, **make a hypothesis**, or educated guess. Using the information you've gathered, try to figure out what the answer to your question might be. Don't get too attached to this hypothesis. It may be wrong.

Fourth, **make observations or do experiments** to determine if your hypothesis is correct. What could you do to prove (or disprove) your hypothesis?

Now that you've completed your experiment, the final steps are to **analyze the data** and **draw conclusions** from it. What did you find out, and what can you learn from that information?

Drawing conclusions is the end of the scientific method, but it is rarely the end of the scientific process. Often you find that your original hypothesis was not correct and you still don't understand what is happening. In this case, you must make a new hypothesis, design a new experiment, analyze more data, and draw new conclusions.

Alternatively, you can share your results with others in hopes that they will do more experimentation on the topic. The scientific community is very open in this regard. Science is a group endeavor in which knowledge is shared and studied by all.

You will learn *how to teach* your children about the Scientific Method in the next chapter. Before you rush into that, though, it's important to understand some of the intricacies of the Scientific Method itself.

Don't worry, this chapter won't overwhelm you with everything at once. Because your children will be doing canned science experiments at first, it's not critical that they learn how to design an original experiment until they are older. Therefore, let's save that for later on in this book; proper experimental design is fully discussed in Part III.

MORE ON ASKING A QUESTION OR STATING THE PROBLEM

When you are embarking on a voyage of scientific discovery, it's wise not to bite off more than you can chew. One of the reasons you state the question or problem is to make you think about what you're doing so you can narrow the focus. Is this really something that you can find the answer to with one experiment?

Sometimes you find a question that cannot be answered by a single experiment. A classic example of this is the old question of whether intelligence is determined by "nature" (genes) or "nurture" (upbringing). Since there are so many facets of intelligence, it would be impossible to design one single experiment that would test them all. As a result, when scientists research this area, they only test one tiny part of intelligence—reasoning skills, for example. We'll talk more about this in the chapter on designing experiments. For now, think of the purpose as a "reality check"—can you really find the answer to your question with one experiment?

The question or problem—often referred to as the "purpose" in lab write-ups—is written at the beginning of an experiment so that people who are reading a report can decide whether or not they actually want to go on reading. If it's a topic that interests them, they'll read on; otherwise, they'll toss it aside.

If you are doing an experiment from a purchased science curriculum, the purpose can play an even more influential role. In "real" science, you would (ideally) try to take note of every little thing, never knowing what unplanned discoveries lie waiting for you. In purchased science curriculum, however, you typically have a definite outcome in mind and focus on observations that are obviously related to the purpose.

Usually, in a lab write-up, you write the purpose second, right after the date. The most common format begins with an infinitive:

> Purpose: to discover which types of food cause the fastest growth in *Helix aspersa* (snails).

An alternative format uses a complete sentence:

> The purpose of this experiment is to discover which types of food cause the fastest growth in *Helix aspersa* (snails).

As you can see, both formats convey the same information. Yet you'd be surprised how strong of an opinion some science teachers hold on this point of writing style. Sometimes, the purpose of your report is the first and last thing the reader sees. Go with whatever format your customer demands; your focus remains on writing a clear but brief purpose that accurately summarizes your goal.

Here is an example of a poorly written purpose:

> Purpose: The purpose of this experiment is to try out different kinds of foods. Snails usually eat plants so we'll try corn, carrots, tomatoes, or other vegetables as I think of them. We could use grains like wheat or barley; maybe we'll even try out meat products like beef and chicken. Then we'll feed a number of these foods to a bunch of snails. But not too many. The snails, by the way, are called *Helix aspersa* and we'll measure their growth and we'll see who ends up as the fastest and coolest snail dude.

As you can see, this so-called purpose overloads the reader with information. The goal of the experiment has been lost.

If you want an example of how concise you can make a purpose, look through a scientific journal. There, you probably won't even see a purpose or question statement. The purpose has been summed up in the title of the article, for example: "Effects of different foodstuffs on the growth rate of *Helix aspersa*". Now that's concise!

MORE ON RESEARCHING INFORMATION

Before doing an experiment, it's best to see if someone else has done something similar. After you do your research, you will find one of three possibilities:

1. Someone has done the exact same experiment before.
2. Someone has done a similar experiment, but not exactly the same.
3. Nobody has ever tried this.

Let's consider each of these possibilities in turn.

Possibility #1: Someone has done the same experiment before.

If the work has been done before, there's nothing wrong with doing it again. Remember, part of the process of learning science is learning to discover things for yourself.

Now suppose that you are not a student, but a scientist researching new ideas. There is still nothing wrong with repeating someone's experiment. If you get the same result, then you've verified the first researcher's conclusion.

Verification of an experiment is an important part of scientific discovery. There have been plenty of people who have claimed to discover something wonderful and new—but when their experiments couldn't be duplicated, their findings were dismissed as worthless.

The cold fusion fiasco of 1989 is a classic example. Drs. Pons and Fleischmann claimed that they had encountered an unexplained source of heat in a flask left at room temperature. Their dramatic conclusion: the heat was generated by nuclear fusion. Subsequent experiments by other researchers, however, failed to verify this conclusion. It was later postulated that the "heat source" was simply a hot spot caused by not stirring the flask[5]. This would never have come to light had the experiment not been repeated.

---

[5] Karl and I were taking freshman chemistry from Dr. Nate Lewis, one of the leading cold fusion detractors, at the height of the fiasco; this was the hypothesis he proposed after trying dozens of possibilities. Others have since come out with alternative hypotheses; see Bad Science: The Life and Weird Times of Cold Fusion by Gary Taubes (Random House, 1993) for more details. The book is not only a fascinating read, but also a testament to the importance of thorough record keeping.

Many people think that the job of a scientist is trying to discover new things. While that is an important part of science, another fundamental part is disproving imperfect or flawed theories. Redoing others' experiments helps in the process of verifying or disproving a theory.

Possibility #2: Someone has done a similar experiment, but not exactly the same.

Try to use the previous work to help direct your hypothesis in a productive direction, and to help form your conclusions.

For example, suppose you wonder: if you drop a basketball and a tennis ball from a second-story window, which will hit the ground first? Before doing an experiment to discover the answer, you do your research. You don't find anything about basketballs or tennis balls. However, you find that Galileo dropped weights from the top of the Leaning Tower of Pisa. He concluded that objects fall at the same rate, regardless of their weight. You are not dropping cannonballs from a height of over fifty meters, but you can still use Galileo's conclusion as an aid.

Possibility #3: Nobody has ever tried this.

This is exciting news. You will be exploring a brand new question, the very frontier of science.

Although you didn't find the results of similar experiments, your research wasn't necessarily a waste of time. You'll often get clues for your hypothesis, as well as ideas on how to test that hypothesis.

Say that you want to test the intelligence of black-footed ferrets. Since there are so few black-footed ferrets left in the world, no one has done any experiments of this nature before. Researching the animals will still give you valuable clues. For example, you discover that black-footed ferrets are nocturnal animals; this may mean that their brains process information more efficiently at night. When you learn that they are predators, you deduce that they might be good at seeking out small objects.

Even small tidbits of information can be helpful to forming an educated hypothesis.

MORE ON FORMING HYPOTHESES

Most people would say that a hypothesis is a guess. It's better to call it an *educated* guess.

You must make the distinction because otherwise, when asked to form a hypothesis, many children pick an outcome randomly, not caring much whether they're right or wrong. It's not unlike flipping a coin, where people pick "Heads" or "Tails" without a second thought. The whole point of the researching step of the scientific method is to give you a platform on which to base your hypothesis.

Be alert for this common problem: many science curricula on the market will short-circuit your efforts to teach children about hypotheses. Right next to the instructions, they spill the beans and tell the child exactly what will happen as a result. How is the child supposed to form his own hypothesis if he is told the outcome in advance?

Also, when children—or adults, for that matter—are told exactly what **should** happen, they tend to ignore results they don't expect. If something goes "wrong," they tend to redo the experiment. They ignore the lessons they might have learned by examining what they did the first time.

Many scientific discoveries have been made by accident, so resist the temptation to ignore unusual results. See the section on reproducibility in chapter 7 for some examples.

One of the fundamental principles of science is that a hypothesis can be disproved. There have been unethical scientists who have juggled their data to suit their hypothesis. Some people get so attached to their pet ideas that they can't bear to have them proven wrong. This is not a new phenomenon. Aldous Huxley wrote the following in 1948:

> "They simplify, they abstract, they eliminate all that, for their purposes, is irrelevant.... They compel the facts to verify a favorite hypothesis, they consign to the waste paper basket all that, to their mind, falls short of perfection."[6]

☞ Rule to Remember

It is okay to have a hypothesis proven wrong. You can learn as much (sometimes, even more) from disproving your hypothesis as you can from proving it. **That's why it's so vital to let your children come up with their own hypotheses.**

---

[6] Huxley, Aldous, <u>Ape and Essence</u>.

MORE ON ANALYZING DATA

When people hear the word "data" they usually think of numbers garnered during the course of an experiment. Now that *is* one definition of the word, but here is another:

"Factual information, especially information organized for analysis or used to reason or make decisions."[7]

In other words, every observation made as you do an experiment is valuable data, even if it isn't a number.

Another thing to keep in mind during data analysis is that all data—especially numerical data—is prone to error. People often assume that measurements, being the most objective data available, are also the most unblemished. This is not the case. The oft-repeated carpenter's maxim to "measure twice, cut once" indicates how common mismeasurement can be.

There are many sources for error in every experiment containing numerical data. Each piece of equipment you use can contribute a small amount of inaccuracy; if you use several pieces of equipment, the inaccuracies can accumulate quickly. Plus, there are many many ways that human error can contaminate the results of an experiment. A person who's unfamiliar with a particular technique may have performed it incorrectly. A child distracted by a younger sibling might have missed the crucial "aha" moment of the experiment by a few seconds.

As you can see, error analysis is absolutely essential to the work of a scientist. Philosophically, your goal isn't to turn your children into math wizards (high-quality error analysis can quickly become very sophisticated). All the mathematics in the world won't help a scientist if he isn't any good at identifying error sources in the first place. Your goal, then, is to raise your children's "error consciousness." In other words, they should be on alert for—and to keep careful records of—experimental error.

Help them understand, too, the attitudes and ideology of science, which values the error-ridden nature of truth and doesn't proclaim experiments a brilliant success without qualification. It is far more important to identify all the error sources in an experiment than it is to prove the hypothesis.

By watching for potential errors while you collect and analyze your data, your children will be better prepared to deal with unexpected results while drawing conclusions. We will discuss this important topic further in Chapter 6.

---

[7] The American Heritage® Dictionary of the English Language, Fourth Edition, Houghton Mifflin Company, 2000.

CONCLUSIONS AND THEORIES

The first thing to do when drawing a conclusion is to decide whether or not your hypothesis was correct. If it wasn't, is there an alternative hypothesis that would fit the data that you've collected?

When a hypothesis has been verified repeatedly by experiments, it may become either a theory or a law.

The difference between a scientific "theory" and a "law" is subtle. A law may describe isolated behavior (such as gravity) and may be little more than a simple mathematical equation. Theories can be more powerful than laws; a theory sheds light on much. You could write a law describing gravity, but a theory would explain how airplanes fly or how bicycles remain upright. Perhaps someday in the future, there will be a theory to explain everything in the universe. Literally.

Scientific theories and laws carry a thousand times more weight than a mere hypothesis. Even so, they aren't cold facts; they can still be disproved, partially or in whole, at a later date by other experiments. Sir Alfred Jules Ayer said it best:

> "There never comes a point where a theory can be said to be true. The most that one can claim for any theory is that it has shared the successes of all its rivals and that it has passed at least one test which they have failed." [8]

Clearly there is no such thing as eternal truth in the world of science.

This doesn't mean, though, that scientific theories lack credibility. Consider this analogy: each time a new picture is taken of the planet Mars in higher resolution than before, a new level of understanding of Mars is achieved, and the old understanding is often partially invalidated. But a scientist shouldn't throw up his hands and say, "My picture of Mars was one gigantic waste of time, because someday a higher resolution picture of Mars could be taken." Nor does it mean that pictures of Mars should be kept hidden from the general public or from students. Today's pictures of the Red Planet don't tell us everything about Mars, but they are credible and well worth studying.

Final acceptance of a new theory (or law) may only come years, even centuries, after the scientist who first proposed it is dead. A scientist looking for glory in his lifetime should be prepared for the possibility of disappointment. At the same time, he should

---

[8] A.J. Ayer, <u>Philosophy in the Twentieth Century</u>, 1982

have faith that the Scientific Method will lead to the advancement of civilization. Just look at all the electric gadgets in your kitchen!

MOVING ON

Now that you understand the theory behind the Scientific Method, it's time to put it into practice. In the next chapter you will learn how to teach these concepts to your children.

CHAPTER SIX
Incorporating the Scientific Method Into Your Curriculum

Most curricula introduce the scientific method all in one lesson. A child is expected to learn all of the steps—and how to implement them—in a single day. What a huge conceptual leap for children to make! It's far better to introduce the ideas in six stages, over the course of a year or more.

☞ Rule to Remember

Remember that the scientific method is the process scientists use to perform experiments. Teaching the scientific method without using hands-on experiments is like teaching driver's ed without allowing your student behind the wheel of a car. You can talk about the skills all you want, but unless they're practiced in the appropriate context, you won't get very far.

In order for this six-stage approach to work, you will need to use a science curriculum that utilizes hands-on experiments. Use a curriculum that has you doing experiments weekly (at a minimum). For specific recommendations, refer to Appendix V.

Here is the six-stage approach.

STAGE 1: EXPERIMENTS AND OBSERVATIONS

The first stage consists of introducing your child to the idea of doing experiments and making observations. This can start as early as kindergarten.

This stage is very simple: just do experiments and follow the observation suggestions in Chapter 3. As you do experiments with your child, talk about what you are doing. At first, you might make observations about the experiment, telling your child what you see, hear, and so on. You should do this for even the simplest of experiments that you do—even the tried and true baking-soda-and-vinegar demonstration.

Later, you can ask your child questions about the experiment to prod observations:

- "Did the color change?"
- "What do you see?"
- "Do you hear anything?"
- "Do you see any baking soda left over?"
- "What is happening to the vinegar?"

At this stage, you do not want to ask any questions which require forming conclusions: "What do you think is happening?" for example, or "Why do you think it did that?"

Do not move on to Stage Two until your child shows signs of approaching the age of reason (see chapter 1).

STAGE 2: "RANDOM GUESS" HYPOTHESIS AND SIMPLE CONCLUSION

This stage introduces the child to the idea of predicting the outcome of an experiment. Before you begin an experiment, ask your child to guess what will happen.

The most generic question you can ask is, "What do you think will happen?" This is a question that could apply to any experiment. However, it's best to customize your questions to the specific experiment. Here are some examples:

- "Which do you think will grow faster, the zucchini plant or the tomato plant?"
- "Do you think the aluminum foil will sink or float?"
- "Will the sugar dissolve faster in hot water or cold water?"

Your child has not yet been exposed to the idea of stating the purpose of an experiment; specific questions are your way of giving him hints about what the purpose might be.

Specific questions also narrow the number of possible hypotheses for your child to choose from. In this first stage, do not ask your children to justify the reasoning behind their hypothesis. It's okay for his hypothesis be nothing more than a random, haphazard guess.

When the experiment is complete, ask them if they were right. Don't get into detailed analysis; a simple yes or no will suffice.

After doing this for a couple of weeks, you can tell your children that this sort of prediction is called a hypothesis. They will be more likely to understand the definition of the vocabulary word once they are familiar with the concept.

Once your children have mastered the skill of trying to predict outcomes, move on to Stage Three.

STAGE 3: THE EDUCATED HYPOTHESIS

When scientists form a hypothesis, they're hoping to make an educated guess rather than a random one. The purpose of Stage Three is to teach children how to base a hypothesis on facts they already know.

To show your children the value of a little information, try the following activity.

 ACTIVITY: Random Versus Educated Guessing

Part I—random guessing
Make a table similar to the one below for writing down results.

Guessed Exactly	Guessed Close (within 2)	Nowhere Near!

Tell your children that you are thinking of a number between one and ten. Have them guess the number. How close were they? Make a tally mark under the appropriate column. Repeat this experiment ten times.

How many times did your children guess exactly? Is it easy to guess exactly when you play this game?

Part II—educated guessing
Make a table exactly like the table for part I.

Tell your children that when you flip a coin, it usually lands heads up half the time and tails up half the time. If you flip a coin ten times, it might end up landing heads up anywhere from zero to ten times. On average, however, it will land heads up five times. (This is the information on which they are to base their guesses.)

Now announce that you are going to flip the coin ten times. Ask how many of those tosses will be "heads." Have your children guess, then flip the coin ten times. Were they correct? At least close? Or way off? Make a tally mark under the appropriate column. Repeat the experiment at least ten times.

How do the results compare to the results of the first experiment? Does your child see the difference between a random guess and an educated one?

Now that your child understands how much of a difference a little knowledge can make, it's time to introduce the concept of researching. Before you do an experiment, ask your child what she thinks will happen. Now get out a reference book—a science text, an encyclopedia, or any book that might contain the answer—and look, together, to see if you can find facts to support your hypothesis.

Eventually, of course, your child will do the research alone.

Ideally, when you're first starting to research, you'll find experiments with answers that are easy to find in your references. Unfortunately, that isn't always as easy as it sounds. So skim through your experiment book to get ideas, then skim your reference books. Are any of the experiments you remember reading about also mentioned in your references?

> Example: Your science encyclopedia mentions Newton's three laws of motion, including the law of inertia. An object in motion tends to remain in motion; an object at rest tends to remain at rest. You remember seeing several experiments on inertia in your experiment book, so you choose those for this week's science lesson.

Science textbooks are not the only sources of information you can use during this stage. There are a wide variety of science reference books available (see Appendix V). If you don't yet own an extensive reference library, try your local library—or even closer to home. That gardening book sitting on your shelf can make a good reference for plant based experiments. The cookbook sitting in your kitchen is a good reference for experiments centered on baking food (see Appendix IV for kitchen experiments).

When your child is comfortable with researching facts and using them to form an educated hypothesis, move on to Stage Four.

STAGE 4: ANALYZING DATA AND DRAWING CONCLUSIONS

Analysis can be exciting–like getting to eat a meal that you've spent so much time preparing.

For the first three or four experiments that you do in this stage, avoid experiments with numerical data. Choose experiments where all of your observations will be made verbally, rather than with numbers.

For this sort of experiment, you and your children will analyze data by talking about what was observed and what that information could mean. You then tie your discussion into the conclusion with questions:

- Which of your observations seem to prove your hypothesis?
- Or, Which of your observations seem to disprove your hypothesis?
- Based on your observations, do you think that your hypothesis is correct or not?
- If not, what would be an alternative explanation of why things happened the way they did?

For example, a variation on a famous experiment mentioned earlier involves dropping two balls of different sizes. Here's how you might question your children to help them analyze the data.

Mom: "So, I have a tennis ball in my left hand and a basketball in my right. If I drop them from the same height at the same time, which will hit the ground first?" *[Note how Mom asks a focused question here instead of just asking what the children think.]*

Child 1: "The basketball. It's heavier so gravity will pull it to the ground faster."

Child 2: "No, the tennis ball. It's lighter so it can move faster."

Mom: "Well, let's try it and see what happens." (Drops balls)

Child 1: "Hey! I think the tennis ball actually did hit first."

Child 2: "No it didn't, I think you were right, it was the basketball."

Mom: "Hmmm. The balls move so fast that it's kind of hard to keep track of them, isn't it? What other observations could we use besides just looking at the balls?"

Children: "Huh?" *[Remember: confusion is a sure sign that you need to rephrase the question!]*

Mom: "What other senses could we use to observe?"

Child 1: "What about touch? If you dropped one of the balls over my left hand and the other over my right, I would feel which one hit first."

Mom: "Wow! That's creative. But let's hear what your brother has to suggest."

Child 2: "What about listening for when they hit? The basketball sounds different when it hits the floor than the tennis ball does."

Child 1: "Your way is easier, but my way is better. Let's do your way first and then try mine."

(Mom drops the balls again.)

Child 2: "Hey, I think they hit the ground at the same time! I only heard one 'thump'!"

Child 1: "I think you're right. Now let's do things my way and see if that's what it feels like."

NUMERICAL ANALYSIS

When it comes time to introduce numerical analysis, you have to add more questions to your discussion.

For example, suppose you are comparing the growth rates of two sets of plants. At the end of the experiment, you measure the heights of the plants; the final results are shown in the table below.

Plant #	Group A	Group B
1	30 cm	60 cm
2	60 cm	75 cm
3	90 cm	90 cm

With this data in front of you, here are some questions that should be answered:

• What is the average size of plants in group A? (60 cm) How much do the plants in this group vary from the average? (plus or minus 30 cm)

• What is the average size of plants in group B? (75 cm) How much do the plants in this group vary from the average? (plus or minus 15 cm)

• What is the difference between the average size of the plants in group A and the average size of plants in group B? (15 cm)

The answers to questions like the ones posed above will affect the way you and your children think about the outcome of your experiments.

When you first look at the numbers, for example, you see that the average height of plants in group B is 15 cm (that's about six inches, for those not fluent in metrics) taller than that of group A. The obvious conclusion is that the plants in group B grow more vigorously.

But now, look at the data more closely. The plants in group A vary from the average by quite a bit—30 cm, or about a foot. The plants in group B vary from the average by as much as half a foot. Yet the difference between the two averages is only half a foot. Does this really seem like a significant difference when you consider how much variation there is in plant height?

For the first few experiments, your children probably won't be able to understand the significance of these numbers. To help them out, graph the data ranges as shown below. The more the two groups overlap, the more likely it is that there really isn't a significant difference between them.

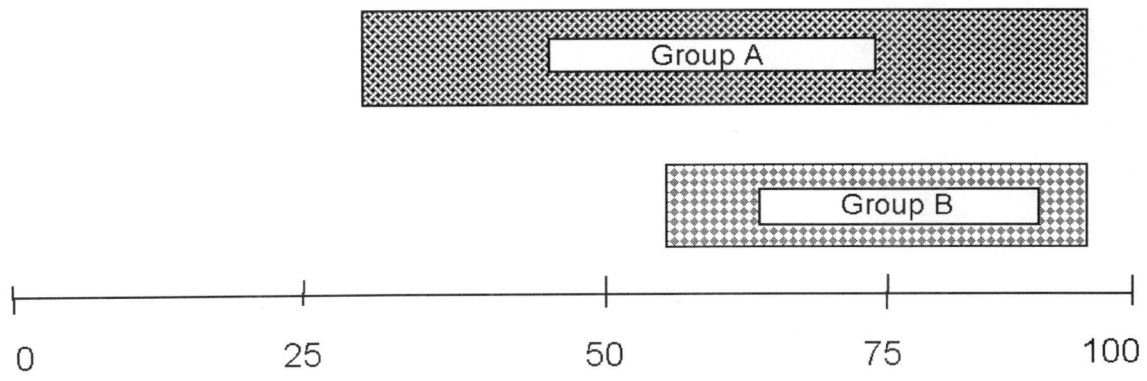

61

In addition to talking about the numbers you collected during your experiment, you also have to discuss whether or not that data may contain errors.

What are some ways error could creep into this particular experiment on plant growth? Maybe you had a difficult time measuring in the first place. If the plant is a large bush, finding the part that's tallest might be a challenge. And when you did find it, how easy was it to run a measuring tape directly to the ground? Maybe there was a large branch in the way that put your measuring tape at a slight angle, thereby corrupting the accuracy of your measurement.

Be sure to discuss the potential for errors even when your data seems to support your hypothesis.

In addition to having the child identify the potential error sources, have him estimate the magnitude of error in his individual data measurements. For example, if a scale is accurate to one ounce, the measurement should be stated as "13 ounces plus or minus one ounce." Sometimes, you can find the accuracy of measuring devices printed in their instruction manuals; other times you just have to estimate it yourself (if so, be conservative and guess on the high side).

Keep in mind that situations involving multiple error sources, or situations requiring complicated math, can require high-school level or college level math skills. But even at this age the child is quite capable of laying the groundwork for such future work by (a) identifying all error sources and (b) estimating their magnitudes. The simplification here is to record the error magnitudes individually—rather than cumulatively or by invoking laws of probability. Of course, this simplification is merely the first and easiest step. But if your children are already in the habit of collecting the necessary lab data, it becomes easier for them to add on the extra layers of mathematical analysis in high school or college.

Your goal is to get them comfortable talking through an experiment, and to expose them to the earliest stages of error analysis. They should be able to draw their own conclusions. If they meet these criteria, move on to Stage Five.

Warning: System Error

As mentioned in Chapter 5, inaccuracies in numerical data can come from many sources. Here are three of the most common:

Limitations of measuring equipment.
For example, if your kitchen scale is marked in one-ounce increments, it might say that something weighs 4 ounces when, in fact, the sample weighs 3.7 ounces. That may not seem like a huge error, but sometimes that small difference in weight can make a huge difference in calculations.

Failings of human judgment.
A casual glance at the ruler shows you that your plant is 4 1/2 inches tall, so you write that down. If you had looked closer, you might have seen that the mark on the ruler was the 1/4-inch mark rather than the 1/2-inch mark. Now you have a faulty data point.

Physical obstructions to obtaining accurate measurements.
Sometimes, it's difficult to measure an object in the first place. If you're trying to determine how much liquid a container can hold, for example, it's better to measure the interior, since that's where the liquid would go. But if your ruler doesn't fit inside the container, you might be forced to measure the exterior instead.

STAGE 5: THE PURPOSE OF IT ALL

In this stage, the way you do science experiments will change—drastically.

Up until now, you have explained what you were going to do in the experiment, then helped your children form a hypothesis by asking them a question. If you remember, the question was meant to sum up the purpose of the experiment for your children.

They now need to learn to figure out the purpose on their own. It's time for your little birds to leave the nest and show that they can fly.

If your children are still not reading very well, continue to read the instructions to them before you begin. If they are good readers, let them read through the instructions themselves. (If your experiment book is one of those that tells what is going to happen, better keep that section covered with a piece of paper.)

Now ask them what you are supposed to find out in this experiment. Since they have spent so long listening to you ask focused questions about the purpose of an experiment, it's quite possible that they will be able to come up with an appropriate purpose statement (or focused question) without further prompting.

If they seem stumped, try using the Socratic Method (see Chapter 2) to help them figure out the purpose.

Say you are doing an experiment on acid rain and plant growth. You begin with two plants. The plants are the same species and are planted in identical pots. The plants are kept together throughout the duration of the experiment. One plant is watered with tap water; the other is watered with a mix of water and vinegar.

Here are some questions that you could ask to help your children understand the purpose:

- At the beginning of the experiment, are the two plants the same or different?
- During the experiment, in what ways do we treat the plants similarly?
- During the experiment, in what ways do we treat the plants differently?

If, after these questions, your children still can't figure it out, ask:

- Do you think the differences in the way we treat the plants will have any effect on them?

Notice that this question is almost, but not quite, focused enough that a child could form a hypothesis from it.

When your children have learned to find the purpose of an experiment without difficulty, you can move on to the last stage.

STAGE 6: THE WHOLE SHEBANG

Now you've laid enough groundwork that you're ready to "introduce" the scientific method. Of course, you've been introducing it all along, but your child didn't realize it!

As shown in the chart in Appendix VI, the steps of the scientific method are:

1. State the Question or Problem
2. Research
3. Make a Hypothesis
4. Make Observations or Do Experiments
5. Analyze the Data
6. Draw Conclusions

Copy the chart in Appendix VI for your child. Also have ready a science experiment that you can do together.

Tell your children that you're going to learn about the methodical process that scientists use to make discoveries. Go through and explain the steps of the scientific method one by one.

None of this information will be new to them. In fact, they will likely find this part of the lesson a bit dry, as it's all review.

Post the chart with the steps of the Scientific Method in an obvious location. You'll be leaving it there at the conclusion of today's experiment, so that every time your children do a science experiment, you can refer to the chart and follow the steps as outlined.

Now do the science experiment, following the steps of the scientific method as you do so. If you've been faithful about practicing the skills in stages one through five, this means just doing things the same way you've been doing them for the last few months.

When the experiment is complete, congratulate your children on having mastered an important part of becoming a scientist.

In Summary

Often, when the Scientific Method is taught, students are simply asked to memorize an acronym. Yet the best way to learn anything is to put it into practice time and time again.

If you follow the six-stage method described in this chapter, the Scientific Method will become an integral part of your children's problem-solving approach. Your children will understand the Scientific Method better than their peers and will be better prepared for more advanced science in the years to come.

CHAPTER SEVEN
Teaching Record Keeping

In order to understand why it is so important for scientists to keep accurate records, it is necessary to comprehend the concept of reproducibility.

Reproducibility means that if one person does something and gets a result, anyone else who does exactly the same thing will get the same result. It's a big word, but the concept is simple enough that even a fairly young child can be led to understand. There's no reason to be afraid to teach your children about reproducibility.

For example, a simple experiment to demonstrate gravity would be to jump off a chair. If I jumped off a chair, gravity would pull me to the ground. In other words, I would fall. Now this law of gravity doesn't just apply to me: anyone in the world who jumps off a chair will fall down.

People trying to reproduce this experiment must follow exactly same procedure that I did. If you jump off the chair, but—unlike me—are attached to the ceiling with a rope and harness, there's a distinct possibility that you may not hit the ground like I did.

Reproducibility is important because the principles of science are true for everyone. When you discover a new principle of science, other people need to try the same experiment and see if they get exactly the same result.

This is where accurate record keeping comes into play. In order to reproduce an experiment, researchers have to be able to do exactly what the original researcher did. If they don't get exactly the same result, they can go back over the records of what the first scientist did (and what they did themselves—they will be keeping records too!) and figure out what went wrong.

Some say that children should write the experimental procedure in the lab notebook before starting the experiment. The idea is to ensure that children understand (or at least have read) the procedure, Unfortunately, if you write the procedure before you begin, you are writing what **should** happen, not what actually **did** happen. It is more akin to fortune-telling than record-keeping.

However, there is nothing wrong with writing down the procedure in advance if you do it on a separate sheet of paper. The point is to teach your children to treat lab notebooks with formality. A professional scientist may need to treat his lab notebook with great formality, even checking it in and out of a secured storage facility, or signing and dating it and getting a witness to do the same. This could be for patent protection or for other reasons.

If you want to convince your child of the importance of accurate record keeping, share some historical stories of things that have been discovered by accident.[9] For example:

- Alexander Fleming didn't mean to have penicillin mold growing on his Petri dish; it just blew in from the open window. This accident lead to the first antibiotics—a major breakthrough in medical science.

- "Gun cotton," the first modern explosive, was discovered when Friederich Schoenbein spilled some acid and wiped it up with his wife's cotton apron. When he hung the apron back up on the stove, it exploded.

If either of these men had not written down what actually happened—or, worse, if they'd discarded the experiments as "ruined"—it would have had serious repercussions on the advance of technology.

In summary, it's very important to develop good record keeping habits. Remember to describe what you do and what you observe as carefully as possible.

🏠 On the Homefront

Although your children aren't likely to discover anything patentable, it's still important to treat a lab notebook with formality. Let me tell you a story about my college chemistry lab. As I was pouring a carefully measured amount of liquid into my flask, I accidentally spilled some onto the lab bench. At the end of the experiment, I didn't have as much product as I expected. Why not? Because, of course, some of the ingredients didn't make it into the flask! Had I written down only what was supposed to happen, instead of what actually did happen, this would not have been clear. A pre-written procedure will never have a notation reading, "spilled small amount of reagent."

---

[9] For a fun read, check out Accidental Discoveries in Science by Royston M. Roberts (Wiley and Sons, 1989) which gives many more examples than the two listed on this page.

ACTIVITY: The Replication Game[10]

Number of Players:	2
Minimum Age:	8-10 years old
Materials Needed:	LEGO™ blocks, paper, pencil

The best way to illustrate reproducibility is to play a game using building toys. It's easiest to use LEGO™ blocks for this game, but it would also be fun to play with simple wooden blocks, Lincoln Logs™, K'nex™, etc. This is a game requiring two people (either two children or one child and an adult). If you have two children playing this game you should have them alternate positions (builder and copier) every game.

You need to make two piles, about ten pieces each, which are exact duplicates. For example, each might have two red squares, a white square, one yellow rectangle, and so on. It is important that the piles be **exactly** the same.

You'll also need to put up some sort of barrier so that one person can work without being spied upon. A large cardboard box tipped on its side works pretty well.

One person is the builder. He constructs some sort of creation out of the sight of the other person (the "copier"[11]). As he does so, he explains what he is doing, using words only. So he might say, "I am starting with the yellow rectangle on bottom. Now I am putting the red square on top of that, right in the middle."

Meanwhile, the copier tries to build the exact same creation by following the directions she is given. The builder may not point ("Put the red square there!") or make any other gestures. When the builder has finished, let the copier compare her model to the original. Are they the same?

The first few times you play, you should let the builder see what the copier is doing. Why? So that the builder can correct any mistakes he sees—orally, of course, as no other help is allowed. As your children improve, however, change things so that the builder can't see what the copier is doing. That forces the builder to think over his choice of words very carefully.

---

[10] The Replication Game is similar to the Description Game in Chapter 3 except that you are describing processes rather than static pictures.

[11] Some people may complain that the word "copier" makes the child sound like a Xerox™ machine. In fact, that's the sort of accuracy we're aiming for in scientific methods. (People who repeat the experiment later need to try to duplicate your methods exactly).

After you reach this point, switch to having the builder build his creation in one room and describe the construction *in writing*. When the copier gets the written instructions, she then attempts to construct his model without any further input from the builder. When she's done, compare the two models. Are there any differences? If so, how could the builder have written clearer instructions?

 On the Homefront

Construction toys are great for beginners because they limit the options for placement. My children, for example, have been known to give instructions like the following: "Place the 1 by 4 red rectangle [LEGO™ block] so that two of the four bumps are on the yellow square, one bump is on the black 1 by 8 rectangle, and the last bump is on the green square." Statements like that make the placement of a piece clear.

Variations

For a more difficult game, try playing with other items where precise placement is harder to describe. Drawing pictures is one option. In this case, you will not describe the finished product; instead you are going to explain how you make each line and curve as you draw them. (e.g. "Begin by placing a circle in the very center of the page. Draw an equilateral triangle underneath it, with the top point resting against the bottom of the circle.")

A fun variation (but one which should only be played occasionally, for obvious reasons) is to frost sugar cookies or cupcakes and decorate them with sprinkles, candies, chocolate chips, etc. See if your children can duplicate each others' cookies.

 ACTIVITY: Practicing Record Keeping

The prerequisite for this activity is mastery of the Replication Game.

When we keep records of an experiment we need to write down not only what we do, but also what we observe. Therefore, your children should be skilled in written descriptions (refer to Chapter 3) before starting this section.

Note: it's not absolutely essential to have your children write in complete sentences when keeping records of experiments. As long as others can make out what happened, it's good enough. After all, sometimes things happen so quickly that it's all you can do just to scribble a few notes down!

One simple starter experiment is mixing vinegar and baking soda to produce carbon dioxide. I love this one because, despite its relatively simplicity, children—and some adults—can do this one several times in a row without getting bored!

Have ready on the table: a bottle of vinegar, a box of baking soda, a clear cup or glass, a stopwatch, and several measuring cups and spoons. You'll also need pencil and paper. Tell your children that they're going to find out what happens when they combines baking soda and vinegar. Caution them to start out with small amounts! Tell them that they need to write down *exactly* what they do and what happens. Tell them to call you when they're done. Then leave the room.

When they're done, tell them you're going to do exactly what they did. Use the same glass (empty the contents first and give it a quick rinse) and follow the directions that they wrote down.

Hopefully they will have left clear notes on their procedure. They should have measured exactly how much of each reagent they added to the glass. If not, prompt them to recognize their mistake—maybe through a series of questions like this: "Did you use two cups of vinegar? No? Then how much did you use? You don't know?" If there's no written record of amounts, they will have to repeat the experiment; be sure you go out of the room while they do so.

When you do your experiment, do your results exactly match the results observed by your children? Of course, you know that both of your experiments will result in the same chemical reaction. The mixture will fizz and bubbles will shoot up from the surface of the vinegar.

However, you don't know how high your children's mixture fizzed unless they tell you so. Did it fizz up to fill half the volume of the glass, or did it overflow?

Also, you don't know how long their mixture went on fizzing unless it's written down. If you time your reaction using a stopwatch, you can ask how that compares to the duration of their reaction. If they didn't write this down, they will probably remember to check next time!

You may want to repeat the baking soda-vinegar experiment a couple of times until your children describe their actions accurately. But before they grow bored of bubbling soda, return to your regular weekly experiment sessions. Each time you perform an experiment, have your children try to record their actions carefully. When you can consistently reproduce their efforts by following their notes, you will know they have mastered this skill at last.

Once they master this skill, have them start keeping records of their experiments in a lab notebook. You don't need to worry about formal lab reports until seventh or eighth grade. Until then, have them record the date, what they did, and what happened.

FORMAL LAB PROCEDURE

Your children should be proficient in following formal laboratory procedure, including doing formal lab write-ups, *before* starting rigorous laboratory courses in high school. Depending on the individual child, you might be able to start as early as seventh grade.

This section explains the protocol for keeping a lab notebook, then proceeds to explain what teachers generally want in a lab write-up. You don't have to introduce these procedures all at once; work your way gradually towards this goal.

THE LAB NOTEBOOK

In college, laboratory work was traditionally recorded in a composition book with graph paper pages. Nowadays, many college labs require carbon-copy notebooks. However, this is as much for the convenience of the instructor (the student tears out the carbon copies and turns them in with the lab report) as anything. For home use, the traditional 4-square-per-inch graph paper composition book is best. A wireless notebook is the next best thing.

Do not use a 3-ring binder. With a binder, you can easily take out pages with experiments you don't like (usually those with data that don't fit your hypothesis). That goes against the whole purpose of record-keeping. The same is true of spiral-bound notebooks; you can remove a page along with any evidence that it ever existed. With a wireless notebook, at least there is a little stub left when a page is removed, showing that something has been taken out.

Always record your findings in pen. This prevents a person, such as yourself, from erasing "undesirable" results. If, by chance, you make an error, cross it out.

What do you write in a laboratory notebook?

First, you write the date, usually in the upper right hand corner of the page. This is a very important step that should not be skipped. In "real" science, every page used is dated.

Second, write a brief summary of what you are trying to do today: the purpose of your experiment. In my "real" (official) lab notebook, this is usually just a few words: "optimizing reagent solution concentrations," for example.

Next you write the procedure along with your observations. Remember, you are not writing down the "ideal" procedure that you are *supposed* to follow; you are writing down what actually happens.

Observations that you make are recorded along with the appropriate step of the procedure.

If you were to separate observations from the procedure, what actually happens would be harder to puzzle out. The following example, in which the observations and procedure have been separated, shows why this is a problem.

The wrong way to record an experiment:

```
Procedure:
I added 10 ml of carbon tetrachloride, then 10 ml of acetone.

Observations:
The mixture turned pale yellow, then a white precipitate formed.
```

When it is written this way, the reader doesn't know when the color change occurred. Did it happen after you added the carbon tetrachloride? Or did it happen after the addition of the acetone?

Here's the right way to do it:

```
I added 10 ml of carbon tetrachloride. The mixture turned pale yellow.

Next I added 10 ml of acetone.  This caused a white precipitate to form.
```

Do you see how much clearer this record is to interpret?

If you are recording numerical measurements, you may want to include a table for your data. For example:

Time	Temperature
0	100°
5 min	90°
10 min	80°
15 min	72°

However, you would not graph the numbers here. That would be analyzing your data, which you will do later.

THE FORMAL LAB WRITE-UP

Most science teachers (high school or college) require you to turn in a formal lab write-up after each experiment. These are a critical part of scientific training, as they serve as a precursor to professional (published) papers.

With this in mind, you might think that there should be one standard for how such reports should be written. Unfortunately, this is not the case, even among professionals: if you were to look at three different scientific journals—*Environmental Science and Technology*, *Applied Geochemistry*, and *Limnography and Oceanography*[12], for example—you would see that all three have slightly different formats. And these are three journals focusing on one field of study, more or less! Journals from different fields of science can vary even more.

The format included in this chapter is representative of what you may be called to turn in to an instructor. However, keep in mind that each instructor has his own pet format for lab write-ups, so if your children ever participate a formal science lab, they may be asked to do things somewhat differently.

Notice that the format for a lab write-up closely parallels the steps of the Scientific Method:

Introduction
An introduction to the experiment. Tells what you want to find out (the purpose). Frequently contains a brief introduction to the theory behind the experiment (your research). Most instructors want the student to keep this short—just a few sentences—but some teachers may expect a whole paragraph or more of background theory.

Procedure
Describe what you did (your experiment) in detail. In your lab notebook, you may have only jotted down a few scribbles or notes on each step; here, you will write things out in complete sentences. Pretend you are writing for a scientific journal.

Data
All data (verbal or numerical) go in this section. Once again, you're going to expand the brief notes from your lab notebook into fleshed-out material. And although you outlined the procedure in the previous section, you may have to

---

[12] Yes, people do actually read things like this. The journals mentioned are three I frequently pull articles from.

refer back to it in this section, in order to make it easy to understand which observations happened when.

Calculations

If your experiment requires numerical analysis, put your calculations or graphs in this section. Explain what you're trying to calculate and how you're doing it. If your experiment doesn't involve numerical data, skip this section.

Results and Discussion

Were the results what you expected? Give your conclusions. If you didn't get the results you expected, was there anything that could have affected your experiment and skewed the outcome? Could you have made errors in measurement or judgment? When you predicted your results at the beginning of the experiment, were there some factors that you forgot to take into account?

In a lab *write-up*, the procedure and data are cleanly separated sections. Why is this okay, considering that it was a big mistake to separate procedure from observations in the lab *notebook*?

There are practical considerations. In the laboratory, you may be frantically scribbling down things into your notebook as fast as the ink can flow from your pen. To use a football analogy, the bodies are flying and the players are shouting. The write-up, on the other hand, is like the post-game show in football; you have a more relaxed situation and you (can) take the time to collect your thoughts.

Furthermore, if someone should develop a question about what happened when, after reading your lab write-up, you can always refer them to your lab notebook as *prima facie* evidence.

IN SUMMARY

Record keeping might seem like a simple enough thing, but it takes time to learn how to do it right, and it takes effort to maintain the necessary rigor. With enough drilling, your children will become adept at it. It's very much a question of habits (good or bad ones).

Record keeping is a vital skill for scientists, but it's also a proficiency that will prove useful in other fields of study. Mastering this art will benefit your children no matter what path they follow in life.

PART THREE

The Well Designed Experiment

CHAPTER EIGHT
An Introduction to Experimental Design

Uh-oh. It's science fair time.

Or maybe you just have budding scientists in the house—your children are not satisfied with doing canned science experiments.

In any case, your children are ready, maybe even eager, to design their own experiments. How should they go about doing it? What contributes to a well-designed experiment?

A well-designed experiment utilizes control groups, double blind procedures (where applicable), a large (enough) sample size, and an intelligently selected sample from a population. Most of all, a scientist should have a very precise idea of what it is he is trying to discover or determine.

If this all sounds like a mouthful, don't panic. This section will explain these ideas, why they are important, and how you can teach them to your children.

You can begin teaching the concepts in this section in the middle school years. Younger children will probably be unable to grasp many of these ideas.

First of all, let's look at some questions that should be asked when setting up an experiment.

WHAT DO YOU WANT TO KNOW?

Choosing a topic may sound like the easiest part of the experiment, but sometimes it's the hardest.

Sometimes children are required to do a science project but aren't given a topic for research. If your children can't come up with a research idea, take the "easy way out": try asking scientists of your acquaintance (relatives, friends, people at church, etc.) for ideas.

If this doesn't work, a gradual narrowing of the field may help. Begin by having your children choose a field of science that interests them the most: biology, for example.

Now narrow that further. If they've chosen biology, are they more interested in animals (zoology) or botany (plants)? Suppose they choose plants. Now they can decide if they want to study one type of plant (e.g. see how "something" affects the plant) or if they want to study several types of plants (e.g. see which species is best at "something").

For those that are still stumped, I've included some deliberately vague science project ideas in Appendix III.

 Sally's Sunflower Study

Here's an example of choosing the topic of an experiment.

Sally, who likes flowers, is trying to come up with an experiment for the upcoming science fair. She knows that her experiment will involve biology, and quickly narrows the field to botany. After all, she would like nothing better than a good excuse to grow flowers.

Now she must choose: would she prefer to grow several types of flowers under the same conditions—for example, should she see what kinds of flowers can survive in her climate? Or would it be better to grow one type of flower under a variety of conditions, figuring out, say, how to maximize flower height?

Sally ponders this question a lot. She loves both zinnias and sunflowers, and she wants to grow both. However, Sally is a good scientist and looks ahead to the end of her experiment. She realizes that zinnias and sunflowers are very different plants and what defines success for one might be radically different from success for the other. Therefore, when the experiment is over, it could be difficult to untangle the effects of climate from the effects of, say, over-watering zinnias. An experiment which compares different species of flowers would best be set aside until Sally has a better understanding of each species of flower on its own.

Most successful science experiments are incremental in nature, and do not leap far beyond the bounds of current knowledge. Therefore, Sally decides to check her ambitions and zoom in on just one type of flower. Sally arbitrarily picks sunflowers. (Perhaps her mother already grows zinnias in the flowerbed out front.)

Now she needs to decide how she will vary conditions among the sunflowers in her test. During the day that she spends considering this problem, she stumbles across an ad in a magazine which proclaims, "Using Acme Plant Food will ensure healthier, more vibrant flowers!" Always a skeptic, Sally decides to test this claim on her sunflowers. By observing the effects of Acme Plant Food (and two of its rivals) on her plants, she will be able to tell whether or not Acme is a good investment.

Sally's sunflower study will be used as an example throughout the remainder of this section.

Is this one experiment or many?

If there are many factors involved, your children may need to do a series of trials rather than a single one.

There is nothing wrong with doing a series of experiments, as long as each experiment of the series is conducted independently of the others. Typically, one experiment is performed at a time; the results of that experiment are used to help form the hypothesis for the next segment of study.

 Sally's Sunflower Study

Sally would like to write a book (or at least a booklet) on procedures to follow for maximal sunflower health. Such a book would have to include amount of water to feed per plant, amount of sunlight necessary per day, soil type, etc. Unfortunately, each of these would require its own experiment—and she only has a limited amount of space available for growing sunflowers! She therefore reluctantly resigns herself to the fact that she will only be able to test one of these factors before science fair time; namely, what fertilizer to use.

Laying Out the Experiment

When your children picked a topic to study, they chose the purpose of the experiment. As you recall from chapter 5, writing a purpose statement helps your children to find the focus of their efforts. The purpose explains what your children want to find out in the end.

In order to understand what tests to do, they must work backwards from this ending point.

 Sally's Sunflower Study

What are the factors Sally should consider when laying out her sunflower experiment? The table below shows the questions she asks herself as she thinks through her project.

Question	Sally's Answers
What did you decide to find out?	Which fertilizer causes sunflowers to grow the tallest?
To find the answer, what measurements will you take? What observations will you make?	I will measure the height of sunflowers which have been fed different types of fertilizer.
What do I need to do before I can make these observations?	Plant sunflowers, feed them with different brands of fertilizer, give them enough growing time that I can be sure of long-term results.
Is there anything I should be careful of, that could adversely affect the results of my experiments?	If I plant the sunflowers too close together, one fertilizer may make its way to the roots of several sunflowers.
What variables might be present in this experiment? How do I limit the number of variables in this experiment to one?	Will be discussed on page 83.
What would my control group for this experiment look like?	Will be discussed on page 83.
What is an appropriate sample size for this experiment?	Will be discussed on page 89.
In what ways could my experimental procedures bias the results of this experiment?	Will be discussed on page 99.

Your children can answer these same questions, though they'll likely get different answers. After doing this, they should be ready to set up their experiment. Be sure that they follow the Scientific Method and keep accurate records as they do their tests.

CHAPTER NINE
Variables and Control Groups

When talking about the scientific method, I mentioned that you should think about the purpose of an experiment to try to limit the scope of your experimentation. If you really want to understand your data, after all, it's best to consider as few factors as possible.

Suppose, for example, that you have a "brown thumb"—you've never met a house plant that you couldn't kill. But you want to have plants in your house, and you think plastic plants look too fake. So you go around asking your friends how they keep their house plants looking luxuriantly green.

You ask three friends and you get three answers. The first friend tells you that she makes sure that her plants get a lot of sunlight. The second friend tells you that she always waters twice a week. The third swears by Acme Plant Food.

You go to the nursery for a new house plant and immediately put the advice into action. You keep the plant in a sunny spot, water it regularly twice a week, and feed it liberal amounts of Acme Plant Food. The plant thrives. It's gorgeous! You get lots of compliments on it.

Then you start to wonder: why is this plant growing so much better than its predecessors did? Is it growing well because of a combination of factors—sunlight, watering, plant food? Or is it just one of those factors that makes a difference?

Does it really need fourteen hours of sunlight to look beautiful? (Fall is approaching, and the days are getting shorter.) Is Acme Plant Food the magic ingredient? (It sure is expensive! Maybe you could cut back?) You are going on vacation for a week—if you skip one watering, will the plant die? Or maybe, just maybe, none of these things make a difference. Maybe you just bought a different type of plant this time, one that happens to be hardier than your previous houseplants.

Sometimes, when doing casual experimentation (as in our houseplant example), we take a shortcut and try many new things simultaneously. But it's best to try just one new thing at a time.

The new thing that's being tried in an experiment is called a **variable**.

At the same time, you should also run another "experiment" which is called a **control group**: a subject (or group of subjects) in which we are changing nothing. This helps us to understand how the experimental subjects react under normal (unchanged) conditions.

Returning to our houseplant example, you could try buying four of the same types of plant at the nursery.

• The first plant you keep the same way you've always kept houseplants. This one is the "control group"—the one that's being raised in a way that you understand.

• The second plant is cared for in exactly the same way as the first, except that it's put in a sunny spot instead of the shady area where the control plant is kept. This plant's variable is *location*—everything else is the same as the control group.

• The third plant is put next to the first plant but is watered more frequently than the other two plants. Here, the variable is *watering*—everything else is exactly the same as the control group.

• The last plant is kept in the shady area and watered on the same schedule as the control plant, but is fed liberal amounts of Acme Plant Food. This plant's variable is *fertilizer*— everything else is the same as the control group.

At the end of one or two months of this treatment, it will be obvious if one or more of the plants is healthier than the others. You will understand which factors contribute to healthy houseplants.

Of course, at the conclusion of this experiment you may be ready to set up a whole new experiment. If regular watering and extra sun both yield healthier plants, what will providing both do? You could set up this experiment with three plants: one with extra sun only, one with extra water only, and one with extra sun and water.

Hopefully you now have an idea of how variables and control groups work.

 ACTIVITY: Too Many Variables Spoil The Broth

A fun way to play around with the concept of variables is to bake things. The "control group" will be made following the recipe exactly as printed in the cookbook. Some variables in recipes are: ingredients (and amounts), baking temperature, and cooking time. You'll have to bake the control group and the experimental group on the same day, so that you can compare the results side by side.

Caution: Don't do this if you can't tolerate the least bit of waste. Most of the experiments will yield edible products, but some won't. If you quadruple the amount of salt in your cookies, even your dog may refuse to eat them.

A simple experiment you can do is to bake something at 25 degrees lower (or higher) than the recipe indicates. Watch it carefully and see how much longer (or shorter) it takes to be cooked through. For this experiment, you can make the batter or dough in one batch and divide it into two portions that will be baked at separate temperatures.

A more complex experiment is to change the amounts of the ingredients to see how it affects the taste or texture. For example, if your cake recipe calls for two eggs, try adding only one egg (or none, or three) and see how the resulting cake compares to the "control group" cake.

For more ideas, see Appendix IV, "A Scientific Tester's Recipe Book."

 ACTIVITY: Green Thumb or Brown Thumb?

Another good way to test out variables is in the garden. Try growing the same crop in three different beds: one grown without any help, one grown using commercial fertilizers, and one grown by an intensive organic method. Weigh the produce from each as it's harvested. At the end of the season, you can tally up production totals and decide which method is the most productive.

Carry the experiment one step further the following gardening season: plant a different crop in the same three beds and use the same methods of gardening that you did last season. (Be sure to keep the organic garden in the original organic bed, and so on.) Again, weigh produce as it's harvested. Is one method consistently more productive than another?

Sally's Sunflower Study

Now let's talk about the variables in Sally's sunflower experiment from chapter eight. What variables might affect our results?

- Growth of sunflowers may be affected by the amount of sun they get each day. Sally wants to limit her experiment to just one variable—effect of fertilizer—so she will have to find some way to eliminate sunlight as a variable. How does she do this? Quite simply, she plants all the sunflowers in areas where they will get equal amounts of sun.

- Different varieties of sunflowers naturally grow to different heights. This variable must be eliminated from the experiment. Sally carefully reads the descriptions printed on the seed packages to ensure that she only plants one variety of sunflower.

Can you think of other variables that might affect the sunflower experiment?

Now that she understands the variables that might affect the growth of our sunflowers, Sally needs to set up her control group. What will her control group look like?

First of all, it will be planted in an area that gets the same amount of sun as the other groups. It will be composed of the same variety of sunflowers as the other group. Since the variable she is testing is "fertilizer," the control group of sunflowers will get no fertilizer whatsoever. This will show her how sunflowers grow without the influence of fertilizer.

IN SUMMARY

If you test too many variables at once, you will not be sure which variable—or which variables—affect the results you get. Try to limit the number of variables in each investigation to one and use control groups. By doing so, you will cement confidence in the conclusions you draw.

CHAPTER TEN
Sampling

When my children hear the word "sample," their minds usually go first to the idea of "free samples." Often, when we are out shopping at a store, merchants give us bite-size portions of their food in order to entice us to buy their wares. The idea is that since one small piece of cookie is so delicious, a large (and expensive) box of cookies would be just as scrumptious.

Imagine, though, what would happen if the merchant used samples that were not representative of what they sold. What if the cookies on the sample tray were fresh, but the cookies in the box that was purchased were a week old, stale and rancid? In this situation, many people would demand their money back. Even worse, they would probably tell all their friends how dishonest the merchant was and caution them not to buy anything at all from him.

As you can see, in merchandising, it is important to have a sample that is representative of the products that are available. The idea of having an appropriate sample is equally important in science.

The laws of science are universal. That is, a law should not apply to just one thing: it must apply to all similar things. And yet it is impossible to test a theory on every single thing in the universe. As a result, when designing an experiment, a scientist must choose a small set of objects to represent "everything."[13]

Let's look at a few principles pertaining to the idea of sampling.

SAMPLE SIZE

When you hear somebody in the news media talking about sample size, they're usually talking about public opinion polls. A poll might say, for example, that 62% of Americans support the president. The polls you see are usually based on surveys

---

[13] For purposes of scientific experimentation, "everything" is referred to as the population. In everyday usage, people use this to describe an entire group of humans. In scientific usage, this could just as well refer to the every member of the species *Mustela nigripes* (black-footed ferrets). Nor is the word population limited in scope to animals; it could just as easily represent every member of the species *Quercus kelloggii* (California black oak) or all esters (a type of organic chemicals).

given to about 1,000 people, who have been chosen to represent the American public in general.

Why one thousand people? Why not ten thousand, or even one hundred thousand? Or why not go the other direction and survey only one hundred people, or even ten?

The problem with polling too few people is that you lose out on accuracy. As an extreme example, what would happen if you had a sample size of one? (In other words, you only interview one person.) The headlines would either read:

100% of Americans Support The President!

or

0% of Americans Support The President!

A sample size of two is much better since the headlines could now report 0%, 50%, or 100% in support of the president—three possible outcomes. A sample size of three is better still (0%, 33.3%, 66.7%, 100%—four possible outcomes). The more people you survey, the more accurate your poll will be.

To be completely accurate, you would have to survey every single person in America. Of course, that would be almost impossible. Even if you could handle the logistics of getting surveys to every person, it's likely that not everyone would respond. (Witness the difficulty the government has running a census every ten years.) Even surveying a number like 100,000 poses great logistical problems and would be too expensive. Therefore, pollsters strive to get opinions from as few people as possible—while still remaining accurate.

For public opinion polls, the happy medium between effort and accuracy seems to be around 1000 people.[14] Next time you see a public opinion poll in the newspaper, read the "fine print" at the bottom of the article; the sample size and margin of error is often given.

How does the concept of sample size relate to science?

Imagine that you want to find the answer to the question, "How long does it take for a green bean seed to germinate?" You put one seed in the soil and wait, and wait, and

---

[14] When determining sample size, professional statisticians don't just pull a number out of the air. They use mathematics to calculate the expected error, given the sample size and certain assumptions. The mathematics behind the curtain could make for an interesting advanced study topic for a high school student, but it is beyond the scope of this book.

wait, and wait…and a year later you are still waiting, because (unbeknownst to you), your seed was a dud!

If you'd planted two seeds, then even if you had a bad one you'd have one germinating seed, which would give you an idea of how long *some* green bean seeds take to grow. If you'd planted ten seeds, then (subtracting the bad one) you'd end up with nine sprouting green beans to give you data points.

Having a large sample size allows you to get rid of the effects of "freak" occurrences such as a bad seed. And of course, with multiple data points, you can take an average to give you a number that's fairly close to the average of most green bean seeds.

A large sample size is especially essential when you're experimenting with humans. One green bean seed is very much like another, but the same cannot be said of two human beings. A three-hundred-pound man and a one-hundred-pound woman may react very differently to the same medication.

The news media routinely reports the results of interesting new medical studies. People sometimes change their behavior because of the outcomes of these studies. And yet some of these tests have a sample size as small as fifteen. Can fifteen people accurately represent the reactions of the entire human race?

No. If the initial medical trial seems promising, scientists normally do a test on a larger group of subjects (hundreds or even thousands of people) before recommending something new to the general public.

With life-science experiments, it's easy to have several specimens undergoing the same experiment simultaneously. In other sciences, you may have to repeat the same experiment several times.

For example, if you are testing how lever length affects how far a catapult will shoot, you will have to shoot the same stone several times at each setting to be confident of your results. If you only did one shot at the three-foot lever length, for example, you might wonder: "Did I release the rope before it was in position? Did I not pull hard enough?" etc. By repeating the process several times, you can remove these doubts from your mind. It's unlikely that you would make the same mistake two or three times in a row.

ACTIVITY: It's Not Easy Being (or Bean?) Green

The seed experiment mentioned earlier is actually a good way to introduce the concept of sample size. You don't actually have to plant the seeds. Just soak them in water for a while, put them in a cup, and put them in a dark place. You'll need to check them at least once a day—not only to see if any have sprouted, but also to moisten them so they don't dry out. Keep track of how many have germinated, and when.

Since you're studying sample size, you'll want to have one cup with just one seed and one cup with several seeds. Compare the results of the single-seed cup with the averaged results of the multiple-seed cup. How close are they? Which result do you trust more?

 ## ACTIVITY: Heads or Tails

You can integrate the introduction of sample size into a math unit on probability. For example, earlier your children were exposed to a simple coin toss experiment. You tossed a coin ten times. Thus, your sample size was ten. The expected result was heads half the time, i.e. five times per experiment. Try the same experiment with a sample size of two, ten, twenty, and thirty coin tosses (and one hundred, if you have the patience). Which of the sample sizes was the closest to the expected number?

 ## ACTIVITY: Merrily We Roll Along

When we throw a die, we expect that each number, one through six, will show up with equal frequency. If you only toss a die twice, however, there's no way that you can have all six numbers show up in equal amounts—you aren't even going to get six results!

With this experiment, you can gather dice from all your board games and throw them all at once. (Theoretically you can also toss multiple pennies at once, but it doesn't work quite as well!) Try sample sizes of six, twelve, thirty, and sixty. Again, compare the results to expectations.

⚲ ACTIVITY: ABC, 123

Another fun way to explore sample size is to try to figure out the most commonly used letters in the English language. Be forewarned that this activity requires time and patience. Children who enjoy cracking codes or playing Hangman should find the results useful, however.

Pick a paragraph in a book, and count how many A's there are, how many B's, how many C's, and so on. Once you've gone through the whole alphabet, you can easily see which letter is most common in that paragraph.

Do the most common letters change if you use a sample size of two paragraphs? How about a whole page?

CHOOSE YOUR SAMPLE CAREFULLY

Now that your children understand the concept of sample size, they also need to understand the importance of choosing their sample carefully.

Unfortunately, it's very easy to choose a sample that doesn't accurately represent the population.

For example, if you did the activity on commonly used letters using a paragraph about zebras in the Zanzibar[15] zoo, you might find that the letter "z" appeared more frequently than it usually does in the English language. Likewise, if your sample page was about quick quails in Qatar[16], you would count an unusually high number of q's.

You can see how results would vary if you weren't careful in selecting subjects for a human experiment. Suppose that your children want to see how the heart is affected by moderate exercise. After conducting trials on their test pool, which is composed entirely of their friends, they come to a conclusion that pulse rates rise by 30% after ten minutes of jogging. Would their results be different if they used, instead, their friends' parents? How about their friends' grandparents?

There are many factors which could affect the outcome of his experiment; age is just one. How regularly does each subject exercise? How much does each subject weigh? Do any subjects have medical conditions?

---

[15] Zanzibar is an island off the coast of Tanzania.
[16] Qatar is a small emirate in the Persian Gulf area.

If they want their results to reflect the human race as a whole, they will have to test subjects who represent different ages, different weights, different levels of fitness, and so on.

In short, if you are doing studies on a diverse group of subjects, be sure that your test group represents the population at large.

Choosing an appropriate sample is related to the idea of limiting variables. Problems arise because many natural populations (humans are the best example) come with "built-in" variables—variables that can be difficult to remove from the equation. Choosing an appropriate sample allows your experiment to represent the whole population, with all its glorious diversity.

☞ **Rule to Remember**

If you can't afford or don't want to test a large diverse population, then consider limiting your experiment to a uniform sub-group. (For example, in the example above, the experiment might be titled, "Effects of Exercise on the Heart Rate of Teenagers.") Although the results of your study won't be widely applicable, at least you will have reduced the number of variables. If the results are promising, you can always perform additional experiments later.

 ACTIVITY: Be Still My Beating Heart

The goal of the activity is to show children why it is important to choose their sample carefully.

The activity requires a total of 8 people. In addition to the parent and child, you need 3 more children and 3 more adults. Children must be under 18 – and the younger the better (although children under 5 won't cooperate). The adults must be over 18 – and the older the better. You also need a stopwatch, pen, and paper.

Take three sheets of paper. Tell your child that the goal is to calculate average pulse rate of human beings. The parent then measures each adult's pulse rate and writes it down on the first sheet of paper. (To measure pulse rate, take 2 fingers and find a throbbing spot on the wrist or neck.) Pulse rate is measured in beats per minute. The parent then measures each child's pulse rate and writes it down on the second sheet of paper. The parent then gives the child 4 numbers "at random" from the entire group (without identifying who the numbers belong to).

Now that you have collected all the data, it is time to analyze it. Ask the child to calculate the average pulse rate of his sample set. What has he learned from random sampling? Probably not much, other than realizing that human pulse rates are not measured in thousands of beats per minute.

Then let him intelligently sample the data. Calculate the adult average pulse rate, and the child average pulse rate. What has he learned now? A significant difference should be apparent between the age groups.

Tip: If you can't detect any difference between the pulse rate of adults and children, the activity is going to fail. If necessary, ask your test subjects jog in place or run for 15 minutes and then measure pulse again. Exercise exacerbates the differences.

Point out to your child how random selection turned out to be less meaningful and insightful than carefully chosen sample sets. In this activity, the random set gives an unpredictable answer that may not be confirmed if the experiment is repeated. By sorting the data by age, results are easier to duplicate—and therefore scientifically more useful.

Question: What is necessary in order to make a randomly selected set give a reproducible result?

Answer: a very large data set, lots of humans tested. A large data set is always an effective strategy—but it is expensive. You probably had trouble enough rounding up 8 people!

❀ Sally's Sunflower Study

Let's apply what we learned in this chapter to Sally's sunflower experiment. What is the smallest sample size she can use that will still give her a quality result?

There are plenty of ways that her sunflowers may be declared "unfit for study." Earlier in the chapter, we mentioned the idea of a seed not sprouting; that could happen just as easily to a sunflower seed as a green bean seed. A sunflower that's afflicted with disease or pests won't grow to its full potential, and will therefore skew the results of the experiment.

If one of her sunflowers is affected in this fashion, she will not want to include it in her final results. Having at least two sunflowers in each grouping helps ensure that she will get at least one data point for that group. However, the more sunflowers she grows in each group, the more accurate her results will be.

Consider, then, growing 10,000 sunflowers. Would she get accurate data with this sample size? Of course! But wait—Sally's yard (and probably yours too!) is just a little too small to do that! In this case, her sample size is limited by growing space. Sally gets her first introduction to a necessary skill that is rarely mentioned in science textbooks: knowing how to make best use of a limited budget.

One way Sally could determine her sample size is based on resources, in her case available land:

1. She begins by figuring out how much growing space she has available. In her backyard, the answer is 100 square feet.

2. She then figures out how much growing space each sunflower needs. We'll say this number is 4 square feet.

3. She divides the answer for question #1 by the answer for question #2. This gives her the number of sunflowers she can grow. In this case, it's 25 sunflowers.

4. Now she calculates the number of test groups by taking the number of fertilizers that are being tested (let's say it's three) and adding one—remember, she has a control group that gets no fertilizer. This gives her the number of groups that she is going to be testing, four.

5. Now she divides the answer for question #3 by the number she found in question #4, and rounds down. Twenty-five sunflowers divided by four

groups equals 6.25 sunflowers. That's how many sunflowers she could theoretically put in each group. But since she can't grow quarter-sunflowers, she has to put six per group.

Even if Sally's resources were unlimited, she wouldn't necessarily want to grow a huge amount of sunflowers. In this case, she might ask herself: "What is the smallest sample size that will still give me a meaningful result, one I—and other people—can have confidence in?" Asking the question is about as far as children can go down this road, since the power of statistical mathematics is beyond their grasp.

With many home science experiments, sample size will be limited by resources available. If your children do experiments on plants, they'll be limited by growing space. If they do experiments on animals, they'll be limited by the amount of cage space they have (or by the amount of money they have for buying animals). If they do experiments with humans, they'll be limited by the number of volunteers they can find.

If it makes your children feel better, you can tell them that professional researchers run into the same limitations. They do small trials at first because that's all they are given funding for. Only after the initial trial has proven successful are they given additional funding to conduct a larger, more scientific study.

In Summary

In this chapter you've learned the significance of performing experiments on large numbers of subjects. You've also discovered how important it is to choose a sample that is representative of the population you are testing. By doing these two things, you ensure that your results will not just characterize the group in your study; they will represent a larger group, the population that you set out to study.

A representative sample ensures that your results will be useful not only to you, but to others. That's vitally important to a scientist.

CHAPTER ELEVEN
Eliminating Bias

Most people have encountered bias at some point in their lives. Unfortunately, by the time a child reaches middle school age, it's likely that at some point they will have found bias directed against them at some point. Whether this is because of some physical characteristic (e.g. skin color or physical disability) or some less tangible characteristic (e.g. being homeschoolers or disliking sports) is immaterial; anyone who has been on the receiving end of bias has learned that it is not a good thing.

It's vital that a scientist perform experiments without bias. This is not as easy as it might sound.

It's not just a question of good intentions. It's possible for even the most "cool headed" and well-intentioned person to unknowingly sway the results of an experiment. Good scientists never consider themselves immune to bias, and are constantly on the lookout for it.

Each and every profession has something that makes it noble. The quest to save lives can make a doctor noble. The quest for justice can make a lawyer noble. The quest to prevent the spread of disease can make a sewer worker noble.

What makes a scientist noble is the quest for truth. Of course, just like doctors sometimes fail to save lives, sometimes scientists fail to find the truth. But in teaching your children to recognize and fight against experimental bias, you are teaching them the very essence of science. This is important stuff.

Now what, exactly, are some ways bias can creep in?

Say you have developed a new brand of mouse food and are testing to see if it produces healthier mice. You design an experiment using two groups of mice. Group A is fed conventional mouse food; group B is fed your experimental mouse food. Your expectation is that your mouse food will produce healthier and heavier mice. As a result, you take the mice in group B out of their cages more frequently to weigh them; you want to track how quickly they're gaining weight. As you carry them to the scale, you pet them and talk to them. At the end of four weeks, the mice in group B are, indeed, heavier and healthier than those in group A.

However, you must ask yourself: is this the result of the food, or is it because the mice in group B received more attention?

A better way to design the experiment is to have a second person involved. This second person would feed the mice the appropriate food. This assistant would know that group A and group B were receiving different food, but not which group was receiving your special food. As a result, the two groups would probably receive equal attention and observation.

In human experimentation, bias may be caused by the subjects themselves. In medical trials, for example, subjects in the control group sometimes show remarkable improvement even though they are only receiving a sugar pill. They think they are getting a remarkable new medicine, and their body adapts accordingly. This phenomenon is known as the placebo effect.

Scientists must know if improvement in experimental subjects is due to the placebo effect or the experiment they are performing. To eliminate bias on the part of the subjects, they do not tell them whether they are in the control group or the experimental group.

Furthermore, professional scientists try to do experiments using double-blind procedures. This means that the scientists observing the subjects do not know which group the subjects are in, because scientists could also unconsciously affect the results of the experiment. Likewise, the subjects are not told whether they are in the experimental group or the control group.

Now that you understand what bias is, and some of the known ways of avoiding it, how do you teach your children about it?

The first two activities are geared to helping children understand what bias is. Bias can only be eliminated if it is recognized. Taste tests make for wonderful examples of the placebo effect.

 ACTIVITY: How Sweet It Is

When your children aren't watching, bake a batch of cookies. Divide them into two plates. When you serve them the cookies, ask which plate has the sweeter cookies. Insist that you added extra sugar to the cookies on the second plate.

What you don't tell them is that you only added one extra grain of sugar.

Alternatively, make one batch with an extra teaspoon of sugar. But make a label that says, "cookies with extra sugar," and apply it to the plate that actually has *less* sugar. Then ask them which cookies taste sweeter.

Chances are that they will pick one plate or the other. They expect that one plate will be sweeter, so their brains adapt their observations accordingly. After they have made their choices, explain the situation and what it is you are trying to teach them.

 ## ACTIVITY: Blue Bread

Make a batch of bread dough. Divide it in half. Add blue food coloring to one portion[17]; leave the other portion with its natural coloring. Shape each portion into a loaf, then let rise and bake as directed in the recipe.

Let your child eat a piece of each loaf of bread. Which does he think tastes better? Chances are that he'll prefer the normally-colored bread. This is not because it actually tastes better—both loaves of bread are made from the same recipe—but because his brain expects the "odd-colored" loaf to taste odd.

Now conduct the experiment using a blind taste test. Blindfold your child and let him taste a piece of each bread. Can he tell which is which? Does one taste better than the other?

While this procedure is less biased than the first—your child can't actually see which bread is brown and which is blue—it's possible that he might pick up unintended cues from you (e.g. your tone of voice). Therefore, this is still not a very good procedure.

If you can get a third person involved, do a double-blind test. To do this, you'll need two blindfolds: one for you and one for your child. Tell your child that you are going to give him the first piece of bread. The third person will hand you a piece of bread, which you feed to your child. Repeat for the second piece of bread. Once again, ask your child which tasted better.

In this procedure, your child has no idea which color of bread he is eating. He cannot see the color, nor can he guess from your tone of voice. Now, hopefully, he will understand why double-blind experiments are preferred in the scientific community.

---

[17] Theoretically, you could use any color for this experiment. People do tend to react more strongly to blue food, however. This may be because there are no naturally occurring blue foods, but more likely it is because blue is the color of most molds.

♀ ACTIVITY: Taste Test

This activity requires three brands or types of one particular kind of food. You might try three different brands of animal crackers, for example, or three different varieties of apples. You can repeat this activity several times, using a different type of food each time.

It's best if you, the parent, serve as tester the first time through. If you repeat the activity later, have your child serve as tester.

Cut or break the foods into bite-sized pieces. Place each type in a separate bowl or bag. Your containers should be labeled only with a letter: "A", "B", "C". On a separate piece of paper, which you'll want to keep hidden until the end of the experiment, write which brand or type of food is in each bowl, e.g.: "A = Fuji Apple, B = Pink Lady Apple, C= Gala Apple."

The next step is to make a randomizer chart; see the sample chart on the next page. The chart determines in what order you will give out the samples. You don't want all of your subjects trying "A" first, all of your subjects trying "B" second, and all of your subjects trying "C" third. This keeps the test subjects blind to what they are tasting, and ensures that no food brand gains an unfair advantage from being tasted first (or last).

Put a blindfold on your first subject. As you pick up the appropriate sample, say, "This is number one." Place the sample in your subject's mouth. Allow him to swallow and take a drink of water before continuing to sample number two. When he has tried all three samples, ask which was his favorite. Indicate this in some fashion on your chart.

Now it's time to do the unscientific or "poorly designed" version of the taste test. Remove his blindfold and have him taste all samples with his eyes open. First taste A, then taste B, then taste C. Tell him what he is tasting, and let the other test subjects listen in. Again, record his preference on the chart.

Now move on to your second subject. Once again, you will put on the blindfold and feed her samples "number one, number two, number three"—even though you are giving her the lettered samples in a different order (perhaps B, C, A instead of A, B, C). Continue as you did for your first subject.

Repeat these steps for each test subject until everyone has had an opportunity to try all three foods. Tally the results. Was there a clear winner in the blind taste test? How about in the regular taste test? Were the results the same in both cases?

The goal of this activity was to demonstrate that different testing methods can lead to different results – even though the same food was being tasted. You are more likely to get good results from this activity if you have multiple test subjects.

SUBJECT #	BLIND TASTE TEST			REG. TASTE TEST
	#1	#2	#3	
1	(A)	B	C	A
2	C	A	(B)	C
3	(B)	C	A	B
4	A	C	B	
5	C	B	A	
6	B	A	C	
7	A	B	C	
8	C	A	B	
9	B	C	A	
10	A	C	B	

Sample chart for Taste Test Activity. This person has already interviewed his first three subjects.

Notice how the samples in the blind taste test are given out in a different order each time.

103

Sally's Sunflower Study

Conveniently, the sunflowers in Sally's won't be able to figure out which fertilizer they're being given, so she does not have to worry about the placebo effect in this experiment.

However, Sally will have already picked out a "favorite" group of sunflowers—one that she thinks will thrive more than the others. She has to, since she's supposed to make a hypothesis about which will grow best before beginning the experiment.

Is there any possible way for bias to creep into her experiment? Of course, there are many.

One example: she could unconsciously water her favorite group more than the other groups of sunflowers. To prevent this, and to give her final result much more credibility, Sally should measure the amount of water she gives each sunflower.

IN SUMMARY

Bias is an important variable in many situations. How a researcher treats subjects—and how a subject treats the research—can have a dramatic impact on the outcome of an experiment. Eliminating bias allows a scientist to gain a better understanding of his results, and therefore come closer to understanding the truth.

PART FOUR

Your Science Education Plan

THE EARLY YEARS

Infant/Toddler (Birth to age 3)

PHILOSOPHY

Why should you even think about science education when your children are so young? First of all, it helps you—the parent—get into the habit of educating children through everyday activities. It accustoms both you and your children to working together and communicating with each other.

In addition, these activities train your children to use their senses to the utmost. They are learning to observe things when they're still wide-eyed with wonder. It is much easier to train your child to observe carefully when they're excited about what they see; the older they get, the less exciting the world in general will be to them.

SCIENCE SKILLS

Observation: Parent makes observations to the child. See page 28. Parent should remain aware of the description categories (Appendix I) and try to familiarize child with each throughout the course of the early years.

Classification: Parent leads child in simple sorting activities. See page 10. Many of these can be incorporated into play or other pursuits integral to the everyday life of a toddler.

Other Science Skills: Not applicable to this age.

What you'll need: Nothing but patience.

SCIENCE CURRICULUM

Textbooks, workbooks, and the like won't work for children this age. It's hard to even read through a simple science book without your child getting wiggly.

Fortunately, at this age, education is a relaxed affair. With this book to guide you, you can come up with age-appropriate science activities. Many activities can be done around the house, and you can go on nature walks and visit zoos and farms to expand your children's horizons.

🏠 On the Homefront

The role of videos (or television) in education is a topic of concern to some parents. In general and prior to age 3 especially, I tried to give my children their science education via games, playful exercise, or hands-on activities, rather than via passive edu-tainment.

The American Academy of Pediatrics recommends no screen time for children under the age of 2. This is not to say that I never once let my children watch a video or TV show at that age along with their brothers, but I tried hard to restrict their viewing time.

Preschool (Age 3 to 5)

PHILOSOPHY

At this phase, you are not worried about teaching facts: you are concerned with training your children's thinking processes. Verbal observations help train them to put their thoughts into words; questions force them to think about what they see; sorting and patterning help them to understand connections.

Above all, make sure that your children are enjoying themselves. If their love of science is quashed this early, it may be hard to recover. If you notice your children showing signs of discontent, by all means, change your approach. For example:

- If you've been working daily, perhaps that's too frequent; cut back to several times a week. (If several times a week isn't enough for a budding scientist, by all means do activities daily.)

- If you've pushed ahead to a new stage and a child begins complaining, you may have leaped ahead too soon—go back to the previous stage for a couple of weeks (or months) before trying again.

Pay attention to your children's verbal and physical cues—they may not use words to tell you that they dislike something. This is a stage of development where a parent has to really listen and pay attention to understand how a child is feeling. That's just as true for science education as it is in other arenas.

If you follow these guidelines, your children will begin to anticipate science time. They'll have a good foundation laid for the serious work that will begin in elementary school. Best of all, they'll get some enjoyable bonding time with their mom or dad.

SCIENCE SKILLS

Observation: Children begin making observations on their own. At this stage, observation is usually limited to objects and is expressed verbally. (See page 28.) Once children are comfortable making verbal observation, parents should begin gentle prompting to help them consider details they might otherwise miss.

Classification: Children learn more advanced sorting skills and are introduced to patterns. (See pages 11-13, 37)

<u>Other Science Skills:</u> Not applicable to this age.

<u>What you'll need:</u> Even more patience!

Both science skills important at this age, observation and classification, are easily worked into activities that your children will enjoy:

- The nature walks that became a staple of your life in the toddler years should continue, giving you and your child ample time to practice observation.
- Sorting can be integrated into both work (helping Mom put away dishes) and play (putting red cars in one "garage" and blue cars in another; having the dolls at the tea party arranged by hair color).
- Once your child has figured out the basic idea of completing a pattern, the two of you can practice with a "game." First, lay out a pattern for your child to complete; then let your child lay out a pattern for you. Naturally, your answer should be correct the majority of the time. However, completing the pattern wrongly is a good way of double-checking your child's patterning mastery. Giggles are usually the first sign that your mistake has been noticed.

SCIENCE CURRICULUM

You're still not going to have a formal curriculum at this age. You can, however, begin reading easy science books aloud: a book you can finish in five minutes or less is ideal, unless you have unusually patient children.

At this early age, it's not important to try and "balance" your children's education by reading equally about all fields of science; instead, read about whatever interests them. You'll find that children generally go through stages of being obsessed with one topic or another—the actual topics will vary from child to child. This is normal and perfectly healthy from a developmental point of view.

🏠 On the Homefront

The role of videos (or television) in education is a topic of concern to some parents. Recall that early in life, my goal was minimizing the amount of screen time, which effectively ruled out the use of video material for the purposes of science education.

By age 3, rather than simply abstaining from video material entirely, my concern begins to shift to the question of quality. Can video help with science education at this age? I don't know if that specific question has ever been studied, but Dr. Jane Healy [*Endangered Minds: Why Our Children Don't Think. New York: Touchstone, 1990*] has written against "jumpy" educational shows (some of which can even be found on PBS) and cites a study where kids whose parents encouraged them to watch such a show had *worse* vocabulary scores than children who didn't watch.

I did find some videos (but not television shows) that I considered educational for this age range. For example, the "how such and such works" type of videos; videos based on the real world. Of course, video material should never play a major role in homeschooling at any age. It is a supplemental tool at best.

EARLY ELEMENTARY EDUCATION
(Ages 5 to 8)

PHILOSOPHY

Children at this age still aren't ready for intensive deductive thinking, so they aren't yet ready to move on to advanced science skills. Focus on mastering observation and classification, as well as introducing the beginning stages of the Scientific Method.

SCIENCE SKILLS

Observation: Child advances to more complex verbal observation and begins simple written observations (see page 30). If child masters these basic skills, methodical observation (see page 31) and observing change (see page 33) may be introduced.

Classification: Child is introduced to scientific classification (see vegetable classification, page 40; car classification, page 41; and airplane classification, page 42) and works with more complex patterns (see page 11).

Scientific Method: Child masters Stage 1, doing experiments (page 55). Once child reaches the age of reason, introduce Stage 2, Random Guess Hypothesis and Simple Conclusion (page 56).

Other Science Skills: Not applicable to this age.

What you'll need: Library card, notebook for nature journal, book of simple science experiments, at least one field guide, and (optional) observational tools such as magnifying glass, tape measure, and spring scale.

SCIENCE CURRICULUM

Your curriculum should be divided into two parts: learning facts and strengthening observational skills.

Learning Facts
To learn facts, get a book or two from the library on each subject and read them aloud. I try to get related books (for example, oceans, whales, sharks, and so on). However, if one of my children has a question, or has a strong interest in a subject, we get books about that also.

<u>Strengthening observational skills</u>
Continue your nature walks. The older children get, the more you should prompt them to describe objects. Second, introduce your children to observing processes. For this, you'll need to do hands-on experiments.

SCHEDULE

Each week have one day where you sit down together and read science books. If your children are independent readers, you could have them read these by themselves. Whether you read to them or let them read by themselves, it's a good idea to discuss the book afterward. This could be done by asking questions about what's been read or by having them doing a Charlotte-Mason-style narration[18].

You'll spend another day each week on the observational activities. You could alternate nature walks one week with experiments the next. In real life, we tend to do more experiments during the winter, when inclement weather is more likely to keep us inside. Then, we spend much of the spring doing nature walks.

For each of these "science days," about half an hour of working with your child is optimal.

---

[18] The ideas of Charlotte Mason, a 19th Century educator, are popular among homeschoolers. If you wish to test comprehension as Charlotte Mason did, have your children summarize what they read orally. In later years, you may wish them to give their narration in written form.

 On the Homefront

The role of videos (or television) in education is a topic of concern to some parents. Starting at age 5, some of my children began to appreciate adult science shows such as NOVA, National Geographic, or nature shows, i.e. educational material that the rest of the family was watching. Of course, their attention span didn't always hold them all the way through such material, and there was still a role for videos targeted towards children.

As always, video material shouldn't be the main focus of an educational plan. By the time my child was capable of reading, I was strictly limiting the amount of television and video—even the educational material. The American Academy of Pediatrics recommends no more than 1 to 2 hours of *quality* TV and videos per day. Books remained the mainstay of our science information.

After age 8, I evaluate the educational potential of video and television material much like I would for myself or an adult. Age 8 is probably a bit young for me to have that attitude, but I like to challenge my children.

LATE ELEMENTARY SCHOOL AND MIDDLE SCHOOL
(Ages 9 to 13)

PHILOSOPHY

For this age group you should stress science skills. Most courses of study introduce these during the high school years, but to do so would be a tactical mistake. Here's why.

During high school, a child begins rigorous study of the sciences. For example, prior to the high school years, children learn principles of chemistry (about elements, solutions, and so on). During the high school years, they begin a very mathematical approach to the subject. (Most students don't have the appropriate math background to do this sort of study until they reach high school.) Learning to balance equations, calculate molarity, and so forth can be very overwhelming. As a result, the science skills tend to be pushed to the side as less important; if not learned earlier, they will be entirely neglected.

Teach the science skills in upper elementary and middle school. High school is too late.

Also, in the event you should decide to stop homeschooling during high school, you will rest assured of one enormous accomplishment. As a parent, you completed the most crucial task in teaching science: imparting science skills. This is an especially significant achievement if your children do not plan a career in science. Just as everyone benefits from basic math skills (but not everyone benefits from calculus), everyone benefits from basic science skills (but not everyone benefits from advanced high school science).

SCIENCE SKILLS

Logical Thinking: Parent begins using Socratic Method (chapter 2) with child. Child is introduced to analogies (see page 14).

Observation: Child increases complexity of written observations. (You may want to continue verbal observations for a while until child is a more fluent writer.) Skills of Methodical Observation (see page 31) and Observing Change (see page 33) should be mastered. The Description Game (see page 35) is an appropriate and fun activity for this age level.

<u>Classification:</u> Child delves deeper into scientific classification, especially classification in nature (see page 42).

<u>Accurate Record Keeping:</u> This skill should be mastered before high school. Appropriate activities include the Replication Game (see page 69), Practicing Record Keeping (see page 70), and keeping a lab notebook (see page 73). Children in grades 7 or 8 should do at least one formal lab write up (see page 75).

<u>Following the Scientific Method:</u> Continue to introduce stages when your child is developmentally ready. Usually, the stages introduced during this time period will be Stage 3, the Educated Hypothesis (see page 57); stage 4, Analyzing Data (see page 59); Stage 5, Purpose (see page 63); and Stage 6, The Whole Shebang (see page 64).

<u>Experimental Design:</u> Child should be exposed to all concepts in the experimental design section, usually in grades 6-8. Begin with one or more activities to introduce each concept involved.

- Activities for Variables and Control Groups: Too Many Variables Spoil the Broth (page 84); Green Thumb or Brown Thumb (page 85).
- Activities for Sample Size: It's Not Easy Being (or Bean?) Green, (page 92); Heads or Tails? (page 92); Merrily We Roll Along (page 92); ABC, 123, (page 93).
- Activity for Choosing Sample Carefully: Be Still My Beating Heart (page 94).
- Activities for Eliminating Bias: How Sweet It Is (page 100); Blue Bread (page 101); Taste Test (page 102).

Once your child is familiar with the concepts of experimental design, practical application—that is, doing an original experiment—will help reinforce the ideas.

<u>What you'll need:</u> Library card, two notebooks (one for nature journal and another for lab notebook), as many experiment books as possible, as many field guides as possible, science references (see Appendix V), optional observational aids mentioned for ages 5 to 8, plus balance or scale, thermometer.

SCIENCE CURRICULUM

You'll want to keep up the nature study. Your child needs the reinforcement of his observation and classification skills. However, the balance between nature study and experiments should be skewed towards the experiment side as your child gets older.

Continue reading science books from the library for factual content. At this stage, have your child read them independently. To gauge reading comprehension, either discuss the reading selection with your child as you did in earlier years, or assign a written report.

SCHEDULE

Keep up the schedule you started in the early elementary years. Reading should gradually be expanded from half an hour to an hour; this can be done in more than one session a week, if your children have trouble keeping still.

Continue to have one day a week for nature study or experiments. If you can, convince your children to increase the amount of experiments you do together. If not, keep the minimum of half an hour of experimentation per week.

☞ Rule to Remember

Be absolutely certain to include science skill mastery as part of your experiment time.

CHAPTER THIRTEEN
Results

As you raise and educate your children, you're likely to gather results from many scientific experiments. The most important results you will see, however, will be in your children.

If you teach your children the scientific skills outlined in this book, what will be the results? Here are a few things that you may find:

- They will be more likely to notice the wonders of nature instead of ignoring them.

- They will look at things in depth, rather than giving them a cursory examination.

- They will be eager to understand how and why things work.

- They will approach problems in a methodical fashion.

- If something unusual happens, they will be unlikely to attribute it to coincidence; rather, they will be eager to discover the causes behind the event.

- They will be less likely to make wild guesses, and more likely to research answers.

Teaching science properly enhances the natural wonder and curiosity of a child. May the joy of discovery and understanding be yours many times as your children grow!

Appendices

APPENDIX I: DESCRIPTION CATEGORIES

When describing an object, it is best to include as many aspects about it as possible. As mentioned previously, these description categories are not meant to constitute a rigidly-followed formula. Rather, they are intended to open your eyes to the possibilities of what factors you might be able to observe in an object.

As you look over the list, you will first consider which categories are applicable to the object you or your children are observing. Some things, for example, do not have a smell of any sort.

☞ Rule to Remember

Any category which does not apply can be left out of the description altogether. There is no need for statements such as, "The rock does not smell like anything."

Sometimes children will have to write several sentences for a particular category. In that case, they should start with a sentence that describes the object overall and then add in details later. For example, for the category "color," a multi-colored object might be described like this:

> The T-shirt is aquamarine blue. On the front, it has a design in white, red, and navy blue.

Use the lists on pages 128-129 for the first few lessons—the detailed explanations help children to get a feel for what to write for each category. Once they're familiar with the various categories, post the brief list (Appendix VI) on the wall as a memory helper.

QUALITATIVE OR QUANTITATIVE?

Sometimes it is best to use words, not numbers, in descriptions. For example, while colors can be described using numbers (e.g. wavelength of light or CMYK values), a good scientist would prefer to use easily understood words like "yellow," "green," "red," "blue," etc.

Some categories are best described in quantitative terms (i.e. numbers). For example, while weight can be described using words (e.g. "heavy" or "light"), a good scientist would prefer to make measurements like "239 lbs" or "2.3 grams." Part of the issue here is that while everybody agrees on what "red" means, not everyone agrees on what is "heavy" and what is "light." An object which is heavy to your toddler may be quite light in a teenager's opinion.

If your children choose to describe an object in quantitative terms, have them do so using the international metric system, rather than the old English system of measurement. Metrics are preferred by scientists, and with good reason: many calculations are straightforward using metric units but can become convoluted if attempted with English units. The metric system offers fewer conversions, greater coherence, a set of convenient prefixes (e.g. kilo, micro), and the decimal system (rather than clumsy fractions). Just as English is the language of air traffic control, the metric system is the mathematical language of science.

And the metric system is not just important in science. We live on a planet where every nation except for Liberia, Burma, and the United States has accepted the metric system. Students who do not understand the metric system will be at a competitive disadvantage, not only in the high-tech industry, but in other careers such as multinational business.

☞ **Rule to Remember**

Use the metric system !

Some concepts can be described either in qualitative terms (words) or quantitative terms (numerical measurements). For example, when describing the size of a rectangular box, it's easy to give measurements. On the other hand, when describing an oddly shaped rock, it may be easier to give a verbal assessment of size: "This rock is as big as an egg."

Every time something needs to be described, children will have to make a choice between "words" or "numbers." It is tempting to say that each category must be one or the other. And yet every rule has its exceptions. For example, temperature will almost always be described numerically, but not always. On our nature walks we often encounter the madrone tree, known colloquially as "the refrigerator tree," as the bark is quite cool to the touch. Figuring out how to accurately measure bark temperature using a thermometer would be a science project in itself.

<u>Example</u>

The leaf described on page 29 was also described by an older child, in late elementary school, using the list of description categories. He determined quickly that temperature and composition weren't appropriate, and that tasting would be dangerous (a good choice, since that particular plant is rather poisonous). The kitchen scale wasn't accurate enough to register the weight of the leaf. And yet, despite these difficulties, he still managed to write a decent description.

This leaf is about six inches high (including the stem) and three inches wide. It is shaped like an oval and has a pointy tip. The stem and the side veins look like a tree without leaves. The leaf is a dark green and the stem is a light green. It is very soft and the top is a bit fuzzy. It weighs as much as a feather. It smells like mint.

Physical description categories

Size and/or dimensions (length, width, height)—For young children, this can be as simple as "big" or "small." For older children you can either compare it to something else ("the size of my fist"), or you can get out a ruler and measure it.

Shape or form—Most objects you observe won't have a nice, regular shape like a circle or square. The odder the shape, the more fun it is to describe!

Color—This can be simple (black, white, red, etc.) or complex (brown with a red belly; white with flecks of black, gray, and gold; etc.) If appropriate, you may want to add an observation on *luster*: shiny, matte (dull), etc.

Texture—You can describe the texture that you see (e.g., "covered with small bumps"), the texture that you feel (e.g., "rough"), or both.

Taste—This category is pretty self-explanatory. Make sure your child knows that not everything is to taste.

Smell—This is one of those categories that you can ignore unless the object has an obvious strong odor. Never smell strong chemicals like household cleaners. (Everyday chemicals, like vinegar or fruit juice, are okay to smell.)   Flowers are a common item with a characteristic smell. Leaves of many plants, when crushed, will often provide wonderful descriptive material for this category as well.

Temperature— Only note temperature if you're describing something which is not at ambient (room) temperature. Use a thermometer to get a quantitative measurement if possible.

Weight—This can be either vague ("heavy" or "light") or very precise (get out your scale).

Composition—In other words, what the object is made of. Sometimes (if you're describing a plant, for example) this won't be applicable. The composition of rocks, manmade items, and so forth can usually be described.

Reaction description categories

Change in Size and/or dimensions (length, width, height)—Has the size changed? Have the dimensions changed? If so, how?

Change in Shape—Has the shape changed? If so, how?

Change in Color—Has the color changed? If so, how?

Change in Texture—Has the texture changed? If so, how?

Change in Taste—Has the taste changed? If so, how?

Change in Smell—Has the smell changed? If so, how?

Change in Temperature—Has the temperature changed? If so, how?

Change in Weight—Has the weight changed? If so, how?

Change in Composition—Has the composition of one or more of the components in the reaction changed? If so, how?

Speed of process—How long does the process take to start? How quickly do the changes take place?

Duration of process—How long does it take for the process to happen, from start to finish?

APPENDIX II: EXAMPLES OF USING THE SOCRATIC METHOD

Although the Socratic Method is explained in chapter two, most people understand the process better when they've actually seen it in action. Here are two Socratic dialogues here to help you understand how the process works in real life.

The Four Steps to Socrates outlined earlier help to orient your thinking, but in real life, your conversation may not always be that neat and tidy. While it's possible to identify the four steps in the following conversations, it also becomes clear that flexibility is a vital part of teaching by the Socratic Method.

Adapting your questions to your children's capabilities, assessing where they stand, and figuring out how to get them where they need to go (from a mental standpoint) can prove to be a complicated dance, as the following discussions illustrate.

DIALOGUE #1

This question was originally asked by a five-year-old. Since children of this age are not ready to form conclusions on their own, I instructed the mother to stay in the observations and facts stages only; when it was time to move on, she should explain the conclusions to her child.

Questions & Answers	Comments
Child: "Mom, why is the sky blue?" Mom: "Are there any times when the sky isn't blue?" Child: "I don't know." Mom: "How about at night? Is it blue then?" Child: "No." Mom: "What color is it at night?" Child: "Kind of black." Mom: "Hmm. What about at sunset? Is the sky blue then?" Child: "I don't know. I don't think so." Mom: "What color is it?" Child: "Kind of bluish-brownish-pinkish." Mom: "What about now? Is the sky blue now?" Child: "Yeah." Mom: "Where is the sun? Do you see it?"	Any time you hear "I don't know," it's a clear sign that you're leaping too far ahead of your child—back up and try asking a simpler question.

Child: "It's up there." *(pointing)* Mom: "Where is the sun at sunset?" Child: "Over there, in the trees." Mom: "Okay. Where is the sun at night?" Child: "I don't know. It's gone." Mom: "Can you see it?" Child: "No." Mom: "The sky's color comes from sunlight shining through the air. The color we see depends on where the sun is in the sky. When the sun is over us, like now, it looks blue. When the sun is low in the sky, the sky looks pink. And when we can't see the sun, we don't see any color at all in the sky. That's why the sky is black at night."	At this point, the mother has elicited all the observations and facts she can from a 5 year old. She now has to explain the conclusions to him, since he's too young to figure them out on his own.

If the child had been old enough to draw conclusions, I would have continued the discussion like this:

Mom: "When the sun is directly overhead, the sky is blue. When the sun is low in the sky, the sky is pink. When we can't see the sun, the sky is black. Why do you think the sky is different colors at different times?" Child: "Because the sun is in different places?" Mom: "That's right. And why do you think the position of the sun affects the color of the sky?" Child: "I don't know." Mom: "Hmm, let's see. Do you remember when we were playing with the prism last week?" Child: "Yes." Mom: "What happened when we held it up in front of the window?" Child: "It made rainbows." Mom: "Do you remember why it made rainbows?" Child: "I think it took the light and took it apart into different colors."	Notice how the mother summarizes the information to help her child digest it. Uh oh! It's the infamous "I don't know." Time to give the child a hint. In order to figure out the answer, the child will need to understand some properties of light. Since we didn't bring these facts forward earlier in the conversation, we have to move back to reminding the child of important information.

Mom: "Right. And what light was it taking apart?"

Child: "Huh?"

Mom: "I mean, where did the light come from?"

Child: "The window."

Mom: "The window was glowing and gave off light?"

Child: (laughing) "No it was sunlight."

Mom: "Right! So there are different colors combined into sunlight. And sometimes it gets broken up so we can see individual colors. Now, can you think of a time when you look at sunlight through something and only see one color?"

Child: "That red clear plastic thing makes it look red."

Mom: "The cellophane. Good! How about when you're in the pool and you look up from underwater. What color does the light look?"

Child: "Kind of blue."

Mom: "Is that because the pool water is blue?"

Child: "Yeah."

Mom: "It looks blue when you stand near the pool and look down. But what if you took a cup and filled it full of pool water. Would it look blue then?"

Child: "No, it's clear."

Mom: "So even though it's shining through something clear, it makes the light look blue."

Child: "Are you trying to tell me that's why the sky is blue? Because the sunlight is shining through something? But what is it shining through?"

Mom: "Well, what's between us and the sun?"

Sometimes, when the child seems to misunderstand the question, a little humor is called for!

Once again, the mother summarizes what the child has deduced. In a long dialogue, summaries can be absolutely critical!

The child is starting to draw appropriate conclusions—victory is near at hand!

133

Child: "Space?"	
Mom: "Well, space is empty, so it has to be something else. Think of something that's closer than that."	
Child: "Oh! The atmosphere!"	
Mom: "Right!"	

DIALOGUE #2

Questions & Answers	Comments
Child: "Mom, do penguins have knees?" Mom: "Why do you ask?" Child: "Well, they walk kind of funny, and they seem to be a big fat blob with feet on the bottom." Mom: "I see. Well, what kind of animals are penguins?" Child: "What do you mean?" Mom: "Are they mammals? Reptiles?" Child: "Oh, I see. No they're birds." Mom: "Well, we don't have any penguins here, but we do have a bunch of birds out in the back yard. Why don't we go look at them. Look at their legs. Do they have knees?" Child: "I guess. Their legs bend, but not the same way as ours do." Mom: "But they do have joints in the middle of their legs like us, right?" Child: "Yeah." Mom: "Do you think penguins could walk without knees?" Child: "I don't know. They walk funny anyway. Maybe it's because they don't have knees." Mom: "Can you see my knees?" Child: "Yeah. Well, not really, 'cause you're wearing pants." Mom: "You can't see them. Does that mean I don't have knees?" Child: "No, they're just covered up." Mom: "If penguins did have knees, do you think they might be covered with something?" Child: "I guess they're really fat. The fat keeps them	The mother realizes that she can't ask her child to directly observe penguins, since Antarctica is a long way away. However, she gets the bright idea to observe something similar. Her first question is designed to show her child the parallel. "What do you mean?"—like "I don't know"—is a clue that you're one step ahead of your child and need to rephrase your question to make it understandable. Mom thinks: "Oh great! This is not going the way I wanted!" Obviously, it's time to try a new tactic!

134

warm but it could also hide their knees."

Mom: "So they might have knees, even if we can't see them. Other birds have knees. Do you think penguins do too?"

Child: "Probably."

Mom: "Let's look in the encyclopedia and see if we're right."

This is about as far as you can get with the Socratic Method: not a definitive answer! To verify your conclusion, turn to other sources.

Appendix III: Science Project Ideas

Why doesn't this book supply you with fully described and detailed science projects? For one thing, no one wants to walk into a science fair and see a thousand identical experiments!

More importantly, being too specific also would deprive your children of the opportunity to do research, figure out the best way to test the hypothesis, and so on. Furthermore, having them customize one of these ideas will allow them to truly think of it as "their" science project, rather than one from a book.

Some of the entries here have more than one suggestion; make sure you choose just one!

Gardening-related projects (Botany)

Which variety of _____ (pick your favorite vegetable) is most productive in your area? For example, if you picked tomatoes, you could compare Roma, Brandywine, and Red Cherry.

If you have a limited amount of gardening space, what is the best use of your soil? Using equally-sized plots of land, grow several different vegetables. You could try determining which vegetable produces the greatest amount (weight of vegetables or, for the daring, number of calories) on that plot of land. Or you could try determining which vegetable saves you the most money over grocery store prices.

Which gardening method (e.g. organic or intensive) gives the highest yields?

Zoology Projects

Most of us have a hard time doing experiments in zoology, since you need a fairly large number of animals to do them properly. However, if you have sufficient space, try the following:

Effects of different types of feed (e.g. commercial feed, grains only, grains and greens, etc.) on growth or milk or egg production. If you're studying growth, be sure to start with brand new chicks or babies.

Which is healthier (or more productive, or faster-growing): free-range animals or caged?

Which breed (of a particular type of animal) is better suited to meat, egg, or milk production in your area?

HUMAN BIOLOGY PROJECTS

Choose one thing about the human body that can be measured quantitatively—weight, body temperature, blood pressure, and pulse rate are four common measurements. Now find an internal or external stimulus such as exercise, diet, or hydration. Is your measurement affected by the stimulus?

GEOLOGY/ECOLOGY PROJECTS

How are different minerals common to your area affected by erosion or acid rain?

Which plants are most effective at slowing erosion? Choose plants native to your area, if possible.

HOUSEHOLD PROJECTS

These are good projects for the girl who says: "Why should I learn science? When I grow up, I want to be a stay-at-home mother."

Choose a common household surface (fabric, painted wall, countertop, etc.). Choose a substance that stains that surface (crayon, marker, etc.). Use common household solvents to try to remove the stain. Which solvents are best at removing the stain? Which solvents do the most damage to the surface during stain removal? Given these factors, which would be the best solvent to use in removing this particular stain from this particular surface? Examples of household solvents: water; water with added dish detergent; peanut butter; acetone (nail polish remover); isopropanol (rubbing alcohol); hydrogen peroxide.

What is the ideal ratio of water to fat to flour, in a recipe for biscuits (or choose another baked product recipe: bread, pie crust, etc...) Be very careful to limit your variables in this experiment—whole wheat flour will require very different ratios than white flour, for example.

Appendix IV: A Scientific Tester's Recipe Book

When you're playing around with recipes for scientific purposes, you don't want to use the most complex recipes you know of. At least, not at first. Here are three good starter recipes.

The recipe size has been reduced so that inedible results give a minimum of waste. As a result, you'll sometimes see a recipe calling for one-eighth of a teaspoon. You may have to approximate this by trying to fill one-half of a 1/4-teaspoon measure.

Remember that you should only test one variable at a time! If possible, make a control recipe—one that follows the instructions exactly—at the same time as the experimental recipe.

Shortbread

1/4 pound (4 Tablespoons) unsalted butter
      (If using salted butter, omit salt from recipe)
1/4 cup sugar
1 cup flour
1/8 teaspoon salt
1/8 teaspoon baking powder

Preheat oven to 350°. Cream butter and sugar together. Add flour, salt, and baking powder and combine thoroughly. Roll dough into a circle 1/4" thick and cut into 8 wedges. Place wedges on ungreased baking sheet and bake approximately 20 minutes, or until light brown around the edges.

Suggested experiments with Shortbread

- Experiment with baking powder. Omit the baking powder. After all, not all shortbread recipes call for this ingredient. Or double the baking powder to 1/4 teaspoon.
- Experiment with salt. Omit the salt. Or double the salt to 1/4 teaspoon.
- Experiment with sugar. Increase the sugar to 1/2 cup. Or try reducing the sugar; cut the amount in half to 2 Tablespoons, or leave it out all together.
- Experiment with fat. Most shortbread recipes you'll see specify that you should use butter as the fat. Is this really necessary? Substitute margarine for the butter. Is there a noticeable difference in texture and taste? Alternatively, try substituting oil or shortening for the butter.

Yellow Cake

1 1/2 cups all-purpose flour
1 cup sugar
1 1/2 teaspoons baking powder
3/4 cup milk
1/4 cup cooking oil
1 teaspoon vanilla
1 egg

Preheat oven to 375°. Combine dry ingredients in mixer bowl. Add milk, oil, and vanilla. Beat on low speed until just combined, then turn mixer to high speed and beat until no lumps remain, about 2 minutes. Add egg and beat for another 2 minutes. Pour into greased and floured 9" round pan. Bake 25 to 30 minutes or until a toothpick inserted in the center comes out clean.

Suggested experiments with Yellow Cake

- Experiment with flour. Reduce flour to one cup. Or increase flour to two cups.

- Change the amount of baking powder to anything from none at all to one Tablespoon—more if you dare!

- Experiment with egg. Eliminate the egg. Or try baking with two eggs. Does the change affect the taste, the texture, or both?

- Experiment with pan shape. Double the recipe. Bake half in a 9" round pan and the other half in a 9" square pan. (Measure carefully; you must make sure that you put exactly half of the batter in each pan.) Which cooks faster?

- Eliminate the oil. How does this change the texture and taste?

- Experiment with beating. Combine ingredients as follows: First, mix dry ingredients in mixer bowl. Add milk, oil, and vanilla and egg mix until just moistened, no more. There will probably still be a few lumps in the batter. Bake as directed in original recipe. How does the lack of extra beating affect the texture of the final product?

White Sauce

1 Tablespoon butter or margarine
1 Tablespoon flour
1/2 cup milk

In a medium saucepan, melt butter or margarine. Stir in flour until well combined. Gradually add milk, stirring constantly.   Cook over medium heat until thickened and bubbly.

(Note: Double amounts if planning to use this in a meal.)

Suggested experiments with White Sauce

- Experiment with flour. Decrease flour to 1/2 Tablespoon (1 1/2 teaspoons). Or increase to 2 Tablespoons. How does this affect the taste and texture?

- Substitute cornstarch for the flour. Begin by substituting an equal amount of cornstarch for the flour. How does this affect the texture? Does it affect the taste as well? You can also try varying the amounts of cornstarch you add to the recipe, as you did in the "experiment with flour" section above.

- Experiment with the fat. Use two or three tablespoons of butter or margarine instead of one. How does this affect the final product? Try using oil instead of butter or margarine. Alternatively, make the sauce without any fat at all; stir the flour directly into the milk.

- Experiment with the liquid. Try substituting any liquid you have available for the milk. The sauce will no longer be "white," of course, but it will no doubt be very interesting! Liquids you can try include fruit juice (try the end product as a sauce for chicken or pork) and broth or stock (end product is gravy). You can even try substituting water for the milk. How is the taste affected?

APPENDIX V: RECOMMENDED BOOKS AND CURRICULA

CURRICULUM

A hands-on curriculum provides you with many opportunities to teach science skills in your home. My favorite elementary-level curriculum is <u>Hands on Science vol. 1: Particles in Motion</u> by Liz Brough. This book makes physical science understandable to children as young as six. That's no small feat, and the experiments do a fantastic job of reinforcing the concepts. It is simply outstanding.

If the questioning mode introduced in chapter two doesn't come naturally to you, I recommend <u>Developing Critical Thinking Through Science</u> by Critical Thinking Press. This book will make your life easier, because each hands-on experiment in the series comes with a scripted set of questions for the parent to use.

If you are looking for hands-on science experiments, there are lots of books on the market. Don't miss out on Reader's Digest's <u>How Science Works</u>. The authors supply a generous helping of background information right in the book (for research) but—this is critical—they don't give away what the outcome will be (allowing your child to form a hypothesis). The steps of the experiment are illustrated and easy to follow. Best of all, <u>How Science Works</u> is just one of a series of excellent books. Other books in this series include: <u>How Nature Works</u>, <u>How the Earth Works</u>, and <u>How the Body Works</u>. My children love them.

Have you ever been frustrated by hands-on experiment books because you never seem to have the "household items" they require? Consider one of the kits available on the market. I really like the <u>Home Science Adventures</u> kits by Stratton House. <u>Home Science Adventures</u> has enough science experiments for a full year and they include everything you need—can you say one stop shopping? The instructions are so clear that older children can do most of the experiments without parental aid.

If you want a curriculum to teach your child pattern and relationship skills, my favorite books are the <u>Building Thinking Skills</u> series by Critical Thinking Press. These books aren't skimpy—they are thorough and provide a lot of practice. They also happen to be a lot of fun! Our children really like them.

A good course in critical thinking is <u>Introductory Logic</u> by Douglas Wilson. This curriculum starts with the absolute basics—what is a statement and what isn't? From there, it moves on to analyzing different types of statements. By the time your children finish the course, they will be able to pick apart just about any argument. The explanations are clear enough that parents without a background in critical thinking will be able to teach their children.

REFERENCE MATERIALS—NATURE STUDY

For younger children, I highly recommend the <u>Finder</u> series published by the Nature Study Guild. The format is a pleasure to use. It reminds me of the <u>Choose Your Own Adventure</u>™ books. You are asked a question with two answers (e.g. Does the tree have needles or leaves?) and turn to the page indicated by the answer. There you find another question with two answers to choose from. Continue flipping through the book in this fashion, narrowing down your choices until you identify the tree you're observing. Young children find these small books (about 3" x 5") much less intimidating than a regular field guide.

As your children grow, you'll want to invest in a larger field guide. Start with the Audubon guide for your area. As of this writing, Audubon field guides are available for the following regions: Southeastern States, Southwestern States, New England, Mid-Atlantic States, Pacific Northwest, Rocky Mountain States, California, and Florida. Each of these field guides contains information on habitats, rocks and minerals, weather, sky maps, plants, insects, birds, mammals, reptiles, fish, amphibians, and more. While each section is not as complete as a specialized field guide, the regional guide is compact and makes a good starting point.

As budget permits, you should buy specialized field guides. These books specialize in topics such as rocks, birds, trees, and so on. Though I've purchased many specialized field guides, and found them to be of great value, I still often pack just my regional guide for nature hikes—taking along the whole library would make for a mighty heavy backpack!

REFERENCE MATERIALS—GENERAL

In order for your children to form hypotheses, they need to research information. It's a good idea to invest in some good general science references.

A set of encyclopedias is ideal. If you don't have the money or shelf space for a full set, then you should at least get a good one-volume encyclopedia of science. At our house, the well-worn favorite is the <u>DK Science Encyclopedia</u>.   Kingfisher also publishes an excellent science encyclopedia.

Specialized books for each branch of science will allow your children to "dig deeper" with their research. DK publishes many Eyewitness books focused on more specific science topics.

APPENDIX VI: REPRODUCIBLE CHARTS

Permission is given for the customer to reproduce this page for the education of his or her children.

Physical description categories

Size and/or dimensions

Shape or form

Color

Texture

Taste

Smell

Temperature

Weight

Composition

Reaction description categories

Change in Size and/or dimensions

Change in Shape

Change in Color

Change in Texture

Change in Taste

Change in Smell

Change in Temperature

Change in Weight

Change in Composition

Speed of process

Duration of process

The Scientific Method

State the Question or Problem

Research

Make a Hypothesis

Make Observations or Do Experiments

Analyze the Data

Draw Conclusions

FORMAL LAB WRITE-UP PROCEDURE

Permission is given for the customer to reproduce this page for the education of his or her children.

Introduction or Purpose

Contains the purpose of the experiment. Sometimes contains a brief introduction to the theory of the experiment. Keep it short.

Procedure

Describe what you did in detail. In your lab notebook, you may have only jotted down a few scribbles or notes on each step; here, you will write things out in complete sentences. Pretend you are writing for a scientific journal.

Data

All data (verbal or numerical) go in this section. Once again, expand the brief notes from your lab notebook into fleshed-out material.

Calculations

If your experiment requires numerical analysis, put your calculations or graphs in this section. Explain what you're trying to calculate and how you're doing it. If your experiment doesn't involve numerical data, skip this section.

Results (or Conclusions)

What were your results? Were they what you expected?

Discussion

If you didn't get the results you expected, was there anything that could have affected your experiment and skewed the outcome? Could you have made errors in measurement or judgment? When you predicted your results at the beginning of the experiment, were there some factors that you forgot to take into account?

Index

ABOUT THE AUTHOR

"Why?" is a question that drive many parents crazy, but it happens to be Robin J. Schneider's favorite question. Her favorite statement is, "There has to be a better way to do things!"

Her natural curiosity and drive to improve on the current state of things led her to science at a very young age. She blazed through most of the science books in the children's section of the library by age ten. She also received a good deal of training from her mother, a published author with an interest in early childhood education, and from her father, a professor of chemistry.

Robin studied at the California Institute of Technology and California State University, Los Angeles. She has a bachelor's degree in chemistry and recently returned to graduate school at the Colorado School of Mines, where she is currently working towards a Ph.D. in geochemistry.

In between her undergraduate and graduate careers, Robin spent fifteen years as a stay-at-home mother. During this time, she homeschooled her four sons, learned lots of fun things herself, and wrote this book.

Teaching has been a constant in Robin's life. In addition to teaching her children, she has given workshops at homeschooling conferences nationwide, taught classes at church for a variety of ages, instructed community groups on emergency preparedness, tutored students in math and chemistry, and taught chemistry labs at the college level. She was the recipient of the Colorado School of Mines chemistry department's 2010 Outstanding Graduate Teaching Assistant Award.

***

One could say that Robin's husband Karl actually *is* a rocket scientist.

Or, at least, he was. Karl spent several years at NASA's Jet Propulsion Laboratory, where he collaborated on such projects as the Mars Pathfinder probe.

Karl's interest in science formed during his high school years. He has a bachelor's degree in engineering and applied science from the California Institute of Technology. After leaving NASA, he worked in the field of embedded commercial software for several years before "retiring" to take over as primary homeschooling parent of the Schneider family.

WONDERING ABOUT HIGH SCHOOL?

Many of the skills taught in this book are applicable to the high school years. For more age-appropriate activities, watch for the next book in this series, *Teaching High School Science at Home (Without Being a Rocket Scientist)*.

For the latest news on this book, log on to the Internet and learn more at http://www.homeschoolzoo.com/THSSpreview.shtml.

Made in the USA
Lexington, KY
25 February 2012